AF581721

Sacred Sites, Rituals, and Performances: New Perspective for Religious Tourism Development

Sacred Sites, Rituals, and Performances: New Perspective for Religious Tourism Development

Editor

Kiran Shinde

MDPI • Basel • Beijing • Wuhan • Barcelona • Belgrade • Manchester • Tokyo • Cluj • Tianjin

Editor
Kiran Shinde
La Trobe University
Australia

Editorial Office
MDPI
St. Alban-Anlage 66
4052 Basel, Switzerland

This is a reprint of articles from the Special Issue published online in the open access journal *Religions* (ISSN 2077-1444) (available at: http://www.mdpi.com).

For citation purposes, cite each article independently as indicated on the article page online and as indicated below:

LastName, A.A.; LastName, B.B.; LastName, C.C. Article Title. *Journal Name* **Year**, *Volume Number*, Page Range.

ISBN 978-3-0365-1867-1 (Hbk)
ISBN 978-3-0365-1868-8 (PDF)

Cover image courtesy of Kiran Shinde

Contents

About the Editor

Dr. Kiran Shinde has published extensively on topics related to religious and cultural heritage and tourism, urban planning, destination management etc. In 2020, he published a co-edited a book titled "Religious Tourism and the Environment" (with CABI). In 2019, he led a UNWTO (World Tourism Organisation) project "Buddhist Tourism in Asia: Towards Sustainable Development" covering Buddhist heritage and tourism 16 countries.

Preface to "Sacred Sites, Rituals, and Performances: New Perspective for Religious Tourism Development"

It is now widely accepted that religious tourism encapsulates the essence of contemporary patterns of travel to sacred and religious sites. On one hand it remains firmly rooted in, and carries forward, tenets from pilgrimage traditions and religious practices. On the other, such visitation includes recreational and leisure components that allows visitors (pilgrims, religious tourists, tourists, and the in-betweens) to experience sacred sites in many ways. This special issue of journal Religions invited perspectives on place-stories, rituals, performances that are central to pilgrimage and sacred sites to explain newer forms of religious tourism.

The purpose of this special issue is to explore the potential of rituals and performances in sacred sites in explaining contemporary religious tourism. From demand side, it is necessary to understand what kind of activities visitors engage in in a sacred site; how similar or different are these from their origins often found in traditional pilgrimages; what are their interactions with religious institutions that are often present in sacred sites to offer ritual services for visitors. From supply side, it becomes crucial to ask questions about how traditional hegemonies of religious actors such as priests, temple-managers, clergymen and other social classes of performers are adapting to changing nature of pilgrimage and religious tourism.

Some of the themes explored in this special issue are:

- Changing landscapes and place-stories in sacred sites
- Reworking of rituals and ritual economies in religious tourism
- Transformation and recreation of performances as attractions
- Impacts of religious tourism on religious-cultural fabric of sacred sites
- Modification/ Innovation in rituals for religious tourism needs
- Engagement of visitors in traditional pilgrimage rituals and activities

Essays in this issue illustrate the dynamic nature of religious tourism that is generated by its embeddedness within ecosystems defined by religious practices, rituals and performances. These constituents of religious tourism are specific to socio-spatial and cultural contexts and yet prone to much wider societal and temporal changes. The explanations here are of immense benefit to researchers and practioners alike for a simplified understanding of religious tourism. Readers can find several examples that demonstrate the significance of religious tourism for sustainable development of destinations. These contributions cover a fair amount of diversity across geographies and religious practices including the temples of Japan, Cathedrals in Europe, Buddhist destinations in India, island worship in Taiwan, and so on. A discussion of megaevents and festivals in creating newer forms of religious and spiritual experience provides fresh perspectives on understanding religious tourism.

I thank all contributors for their insightful essays in enriching the emerging scholarship on religious tourism.

Kiran Shinde
Editor

Editorial

Sacred Sites, Rituals, and Performances in the Ecosystem of Religious Tourism

Kiran Shinde

Department of Social Enquiry, School of Humanities, La Trobe University, Melbourne 3086, Australia; K.Shinde@latrobe.edu.au

It is now widely accepted that religious tourism encapsulates the essence of contemporary patterns of travel to sacred and religious sites. On one hand, it remains firmly rooted in, and carries forward, tenets from pilgrimage traditions and religious practices. On the other, such visitations include recreational and leisure components that allow visitors (pilgrims, religious tourists, tourists, and the in betweens) to experience such sacred sites in multiple ways—a trend that has also been observed in many traditional pilgrimage travels. Much of the existing literature, seems to categorize visitors into (sometimes) exclusive and (other times) into overlapping categories based on their motivations. The focus of their journey and the destination is largely determined by the spiritual magnetism of the sacred place. However, what is it that they do in that sacred place that helps to sustain, reinforce, and reproduce its attraction? There should be something more than the place that anchors religious tourism. This is where one must explore the potential of rituals and performances in sacred sites in order to explain contemporary religious tourism.

To explain newer forms of religious tourism, this volume draws attention to different perspectives on place stories, rituals, and performances that are central to pilgrimage and sacred sites. The early literature on pilgrimage studies has been heavily influenced by Turner and Turner's (1978) classic work, where they argued that rituals in pilgrimage, created a rupture and sense of liminality for pilgrims from their everyday life. This essentialist approach, focused on the pilgrim (visitor) and became the bedrock for the debates around the differences between pilgrims and tourists—something that continues to occupy pilgrimage literature. Turnerian tradition seems to have fallen short of giving due credit, to other actors and institutions that assisted the pilgrims in the performance of rituals. Later scholarship strongly argued in favour of the agency held by religious actors, in reinventing rituals for producing and reconstructing the sacredness of a place. The associated consumption economy around rituals also emerged as part of scholarly discussions. Many of these studies were situated in the western context. In comparison, the centrality of both rituals and performances was recognised quite early on, in studies of pilgrimage places in the East and the Orient.

One of the significant conceptualizations to understand pilgrimage and pilgrimage places, based on rituals and tradition, was the Sacred Complex Model. In this model, comprehensively demonstrated by Vidyarthi (1961), a pilgrim-place was seen as developed around a robust ritual economy embedded in religious, as well as cultural, social relationships and networks. This approach puts rituals at its core and comprises three interrelated analytical categories: Sacred geography, sacred specialists, and sacred performances (Vidyarthi 1961). Sacred geography refers to the physical ensemble of landscape, elements attributed with sacred value, based on stories and legends of gods and deities, that are believed to inhabit that place. It is important that the sacredness of the place be known and made accessible to ordinary people. This is where sacred specialists play the most important role in mediating visitor experience. To partake on the sacredness of the place, visitors need to engage in a variety of rituals and practices, which are collectively known as sacred performances. Despite its criticism, the model has been fruitfully applied to unpack

Citation: Shinde, Kiran. 2021. Sacred Sites, Rituals, and Performances in the Ecosystem of Religious Tourism. *Religions* 12: 523. https://doi.org/10.3390/rel12070523

Received: 20 June 2021
Accepted: 7 July 2021
Published: 12 July 2021

Publisher's Note: MDPI stays neutral with regard to jurisdictional claims in published maps and institutional affiliations.

and explain the socio-spatial fabric of pilgrim towns, owing to rituals and performances as well as the patronage relationships that they generate.

The interdependencies between the geography of the place, its rituals, and specialists, is increasingly becoming an arena of scholarly study, in other places as well. Those institutions play a significant role in any setting, as it is forcefully argued by Eade (2020), through a comparative study involving the Catholic shrine of Lourdes, France, and the pre-Christian shrine of Avebury, England. Using the two different settings, Eade (2020) analyses the ways in which institutions gain power and agency in articulating rituals (including ceremonies) for pilgrims and seekers, and how those rituals mediate their interactions with the sacred landscape. Similar emphasis on performances and performers is seen in the influential work of Reader (2007) in the case of Japanese Shinto shrines. It was found that "many Buddhist priests and officials of Buddhist organizations ... emphasized the importance of pilgrimage as a means of bringing people to temples and advocated the creation of new pilgrimage routes to facilitate the process" (Reader 2007, p. 219). In a recent study of pilgrimage in Iran, Moufahim and Lichrou (2019) have convincingly argued that material culture and ritual consumption plays a significant role in "achieving forms of authenticity" in pilgrimages (p. 322). Pilgrims consume the sacred space in many ways, but most importantly, the objects, materials and souvenirs used in such rituals assist to translate the abstract notions of the sacred into a tangible form of doing and experiencing (Higgins and Hamilton 2011). Nevertheless, rituals are dynamic, they can be preserved, adapted, appropriated and invented and re-invented (Laing and Frost 2014).

Do the abovementioned foundational concepts in pilgrimage, help in explaining contemporary religious tourism, which has become eclectic and evolved as a postmodern phenomenon? The answer is partly yes, but such concepts need to be revised and updated, because contemporary religious tourism is more than pilgrimage. Structurally, one can think of a pilgrimage place as a religious setting. According to Kerestetzi (2018), a religious setting is "a coherent space in which objects, bodies, actions, and ideas form a system—or even an ecosystem" (para 2). This also represents "areas of copresence between humans and deities, as well as points where real space and mythical and imagined places meet and intertwine" (Kerestetzi 2018, para 17). While Kerestetzi mentions the concept of ecosystem in fleeting manner, it provides a foundation for developing an approach for religious tourism that is conceptual and pragmatic. The idea of an ecosystem is promising, because it allows for a "systematic and procedural" examination of connections, amongst the constituting parts and their relationship to the whole and finally the whole as one.

The key ideas from the existing literature can be summarised into a conceptual approach of religious tourism, as an ecosystem in which, there is a sacred/divine (inherent by nature or constructed) resource that manifests itself through all material and non-material things, and its experience depends on several factors. These factors, on one side, include the religious framework that defines rituals and performances necessary for a religious and spiritual experience, whereas, on the other, they include the more contemporary touristic dimensions, seen in the consumption of the same setting, for an experience of heritage or the exotic other, while using a tourism infrastructure. In between, there could be multiple combinations possible, where visitors may choose to engage in certain rituals, in alignment with their motivations and the experience they seek. Framing religious tourism as an ecosystem of resources, producers, and consumers, allows for a more comprehensive understanding of religious tourism. There could be further sub-ecosystems depending on what resources are at play. This structural approach is graphically presented in Figure 1.

The articles in this volume explore many dimensions of the demand–supply relationships that are discernible in the proposed ecosystem approach. From the demand side, one can examine what kind of activities visitors engage in at a sacred site; how similar or different are these from their origins often found in traditional pilgrimages; what are their interactions with religious institutions that are often present in sacred sites to offer ritual services for visitors. From the supply side, it becomes crucial to ask questions about how traditional hegemonies of religious actors such as priests, temple-managers, clergymen, and other social classes of performers are adapting to the changing nature of pilgrimage and religious tourism. What are the materials and objects used in rituals and performances and how do they generate different relationships between producers and consumers in experiencing the sacred? Many of these ideas are explored in the collection of essays in this volume as discussed below.

Figure 1. The ecosystem of religious tourism.

The centrality of a sacred resource and its appropriation through the material culture of rituals is ably discussed in two essays. Zang argues that consistent reworking and inventions of rituals at the temple of a popular goddess named Mazu at Meizhou island has increased its religious tourism appeal. Additionally, in a dialectic relation, the increasing religious tourism activities are reinforcing the imagination and presentation of the goddess Mazu as "symbol of common cultural identity in mainland China" for Mazu worshippers. Zang argues that by augmenting many traditional rituals such as "dividing incense", the temple and its authorities have been able to recreate such a strong cultural identity. The transformations in rituals and performances have also influenced, as well as led to many architectural and liturgical reconfigurations that add to the multi-functionality of the temple, as both a sacred site and a tourist attraction. Many rituals have also been appropriated by local and central governments, as heritage products, for promoting tourism to the island and not surprisingly the Chinese diaspora is attracted to the site in huge numbers, further patronizing the new developments at the temple.

The development of a shrine as tourism attraction is ably demonstrated by Mia Tillonen in her study of Seimei Shrine in Japan. Although a humble local level shrine, the popularity of Seimei Shrine grew after it was presented as a setting for the legend of Abe no Seimei (921–1005) in a novel series and a movie. Capitalising on its increasing fame, the shrine began to install themed statues to realize the legend of no Seimei in material form and reinvented many rituals and performances in which visitors could have sensory experiences of touch and remembrances. The need for these experiences has also provided the basis for the consumptive practices of both religiously motivated pilgrims and those seeking touristic experience. Thus, reinvention of rituals allows the shrine to present itself for multifaceted tourism orientation.

Rituals and performances are the means with which worshippers connect with the sacred resource. This connection is fostered by specialists who are sanctioned to do so. Just how dynamic these connections are is explained by Michelangelo Paganopoulos using his ethnographic research at two monasteries in Mount Athos, Greece. Comparing two rival neighbouring monasteries, he shows how monks are reinventing and reappropriating the traditional values of hospitality in catering to the needs of visitors. Most importantly, through performance of traditional rituals, they emphasize authenticity and influence visitors to take on roles as "pilgrims". Thus, rituals provide agency to both hosts and guests as producers and consumers in claiming authentic experiences of pilgrimages.

Through their examination of Buddhist sacred sites, Geary and Shinde, bring the focus back on to religious rituals and traditions as central to pilgrimages. Exploring the emergence of rituals through textual references and drawing on ethnographic research, they talk about the ecology of ritual observances undertaken by most Buddhist pilgrims during their pilgrimage to sites related to Buddha. They argue that the pilgrims connect with the memory of Buddha through the material and corporeal aspects of these rituals. Pilgrims belonging to different paths of Buddhism come from many different countries and because of this they always travel with their own monks and guides from their respective countries, for facilitating their travel and assistance with devotional activities at the pilgrimage sites. Regardless of the diversity of rituals embedded in their own socio-cultural identities, there is a shared sense of place that is evoked because the resource is the same—Buddha.

Izabela Sołjan and Justyna Liro discuss the ways in which a sacred resource is manifested in the physical environment and how it leads to different types of spatial processes in pilgrim-towns, based on a study of twenty-six popular Catholic sanctuaries in Europe. They identify a spatial hierarchy of contact zones where visitors, including pilgrims and tourists, interact in different ways with the sacred core of the town. It is well known that a sacred core develops around the main shrines and constitute the most intense zones for performance of pilgrimage-related rituals and activities. As the intensity of pilgrimage-related functions decreases over space, other configurations emerge. Pursuing this line of thinking, Sołjan and Liro, conceptualize spatial configurations in pilgrim-towns as sanctuary service zone and based on pilgrim-space interactions further identify a hierarchy within such zones: compact zones (which may be linear or clustered as districts with dominating pilgrimage function), dispersed zones (integrated into the urban space, with their pilgrimage function coexisting with other urban functions), and initial (slightly developed) zones. Delineation of such zone depends on several factors including the history of the place, its hierarchical ranking for pilgrimage practice, and its location in relation to the centre of a town or a city.

The changing roles and imaginations of priests (as producers and consumers of sacred experiences) come across persuasively in Ruth Dowson's paper. Using an ethnographic approach, Ruth Dowson has explored pilgrimages undertaken by priests on motorbikes in a new diocese in Yorkshire, UK. The motorbiking priests have formed a community as "Biker Revs" and through their travel they not only connect with local people but also find new meanings of their own roles as clergy as they navigate through different places. The interactions on motorbike journeys are almost like "religious retreats" that reinforce the connectedness of participants with sacred landscapes.

Religious festivals are more direct opportunities to explore the ecosystem linkages. While festivals derive from the sacred resources embedded in a place or a person or an event, their organization and management highly influence the experiences of that religious event for the participants. Religious rituals and performances and rituals are scaled up during religious festivals so that more and more people can participate in those celebrations. Joanna Bik and Andrzej Stasiak examine the initial stage of the celebration of the World Youth Day in 2016, which took place in the Archdiocese of Łódź in Poland. As a major event, the World Youth Day (WYD) takes place over a few weeks, but equally important are the preparatory days or the "Days in Dioceses" that set the tone of the mega-event. These preparatory events allow volunteers and participants to spend a good amount of time at different locations across different cities and parishes; their activities under the guidance of clergy reinforce religious values and traditions of the Church. While such events socially and liturgically bind together the young pilgrims from all over the world they are also integral to the "promotional and image-building part" of the place. Clearly, an event of faith is productively and effectively used by different tiers of government in association with the Church for marketing to Łódź to foreign tourists.

There are multiple layers of consumption in a festival. Kuo-Yan Wang, Azilah Kasim, and Jing Yu examine a religious festival in relation to consumer choices and market segmentation. From their study of the festival of Mazu in Taiwan, they are able to identify and classify visitors in four major categories: Fun traveler, devout believer, cultural enthusiast, and religious pragmatist, each with different sets of motivations and behavioural patterns. With such a classification, the authors claim that festival organisers and managers can better promote festivals to potential visitors for specific experiences. However, what is more important to note is that festivals continue to epitomize the original doctrine of the religion and that comes through religious rituals which retain the essence and spirit of faith of believers. Thus, rituals and performances in religious festivals remain the centrepiece of religious tourism.

The collection of articles in this Special Issue point to new ways of thinking about religious tourism, but without losing the sight of the fact that such contemporary travel, is built on long histories of ritual practices that help to maintain its distinction from other forms of tourism.

Papers in this Special Issue:

Geary, David, and Kiran Shinde. 2021. Buddhist Pilgrimage and the Ritual Ecology of Sacred Sites in the Indo-Gangetic Region. *Religions* 12: 385.

Tillonen, Mia. 2021. Constructing and Contesting the Shrine: Tourist Performances at Seimei Shrine, Kyoto. *Religions* 12: 19.

Paganopoulos, Michelangelo. 2021. Contested Authenticity Anthropological Perspectives of Pilgrimage Tourism on Mount Athos. *Religions* 12: 229.

Sołjan, Izabela, and Justyna Liro. 2021. Religious Tourism's Impact on City Space: Service Zones around Sanctuaries. *Religions* 12: 165.

Dowson, Ruth. 2021. 'Biker Revs' on Pilgrimage: Motorbiking Vicars Visiting Sacred Sites. *Religions* 12: 148.

Bik, Joanna, and Andrzej Stasiak. 2020. World Youth Day 2016 in the Archdiocese of Lodz: An Example of the Eventization of Faith. *Religions* 11: 503.

Wang, Kuo-Yan, Azilah Kasim, and Jing Yu. 2020. Religious Festival Marketing: Distinguishing between Devout Believers and Tourists. *Religions* 11: 413.

Zhang, Yanchao. 2021. Transnational Religious Tourism in Modern China and the Transformation of the Cult of Mazu. *Religions* 12: 221.

Funding: There is no source of funding for this paper.

Conflicts of Interest: The author declares no conflict of interest. The author is the guest editor for the Special Issue.

References

Eade, John. 2020. The Invention of Sacred Places and Rituals: A Comparative Study of Pilgrimage. *Religions* 11: 649. [CrossRef]

Higgins, Leighanne, and Kathy Hamilton. 2011. Sacred places: An exploratory investigation of consuming pilgrimage. In *North American Advances in Consumer Research*. Edited by Darren W. Dahl, Gita V. Johar and Stijn M. J. van Osselaer. Duluth: Association for Consumer Research, vol. 38, pp. 262–67.

Kerestetzi, Katerina. 2018. The spirit of a place: Materiality, spatiality, and feeling in Afro-American religions. *Journal de la Société des américanistes* 104. [CrossRef]

Laing, Jennifer, and Warwick Frost. 2014. *Rituals and Traditional Events in the Modern World*. London and New York: Routledge.

Moufahim, Mona, and Maria Lichrou. 2019. Pilgrimage, consumption and rituals: Spiritual authenticity in a Shia Muslim pilgrimage. *Tourism Management* 70: 322–32. [CrossRef]

Reader, Ian. 2007. Pilgrimage growth in the modern world: Meanings and implications. *Religion* 37: 210–29. [CrossRef]

Turner, Victor, and Edith Turner. 1978. *Image and Pilgrimage in Christian Culture*. New York: Columbia University Press.

Vidyarthi, Lalita Prasad. 1961. *The Sacred Complex in Hindu Gaya*. Bombay: Asia Publishing House.

MDPI

Article

Transnational Religious Tourism in Modern China and the Transformation of the Cult of Mazu

Yanchao Zhang

Department of Philosophy, Xiamen University, Xiamen 361005, China; yczhang1028@xmu.edu.cn

Abstract: This article explores transformations in the worship of popular goddess Mazu as a result of (religious) tourism. In particular, it focuses on the role of transnational tourism in the invention of tradition, folklorization, and commodification of the Mazu cult. Support from the central and local governments and the impact of economic globalization have transformed a traditional pilgrimage site that initially had a local and then national scope into a transnational tourist attraction. More specifically, the ancestral temple of Mazu at Meizhou Island, which was established as the uncontested origin of Mazu's cult during the Song dynasty (960 to 1276), has been reconfigured architecturally and liturgically to function as both a sacred site and a tourist attraction. This reconfiguration has involved the reconstruction of traditional rituals and religious performances for religious tourism to promote the temple as the unadulterated expression of an intangible cultural heritage. The strategic combination of traditional rituals such as "dividing incense" and an innovative ceremony enjoining all devotees of "Mazu all over the world [to] return to mother's home" to worship her have not only consolidated the goddess as a symbol of common cultural identity in mainland China, but also for the preservation of Chinese identity in diaspora. Indeed, Chinese migrants and their descendants are among the increasing numbers of pilgrims/tourists who come to Mazu's ancestral temple seeking to reconnect with their heritage by partaking in authentic traditions. This article examines the spatial and ritual transformations that have re-signified this temple, and by extension, the cult of Mazu, as well as the media through which these transformations have spread transnationally. We will see that (transnational) religious tourism is a key medium.

Keywords: ancestral temple of Mazu; commodification; diaspora; folklorization; invention of tradition; pilgrimage; religious tourism; sacred sites; transnationalism

check for updates

Citation: Zhang, Yanchao. 2021. Transnational Religious Tourism in Modern China and the Transformation of the Cult of Mazu. *Religions* 12: 221. https://doi.org/10.3390/rel12030221

Academic Editors: Kiran Shinde and Knut Axel Jacobsen

Received: 25 January 2021
Accepted: 15 March 2021
Published: 23 March 2021

1. Introduction

This article explores new developments in the worship of Mazu, one of the most popular goddesses in China, due to growing (religious) (trans)national tourism. After contextualizing the cult historically, I will show the contemporary transformations it has experienced, highlighting the actors, media, and processes involved. We shall see that transnational tourism plays a central role in these transformations. To undertake this task, I rely primarily on textual sources to reconstruct how the cult was founded and rendered orthodoxic throughout the centuries. I will juxtapose this with reconstruction accounts of contemporary practices, as well as my own ethnographic work, which included participant observation of important events in the worship of Mazu at her ancestral temple, such as her birthday, the day of "her ascending to Heaven," and Meizhou's Mazu Cultural Tourist Festival.[1] This juxtaposition will throw the transformations into relief. A mix of sources—textual and ethnographic—are necessary to place current dynamics in their longue-durée context.

[1] I did my fieldwork study in May and November 2020, and also participated of Meizhou's Mazu Cultural Tourist Festival in November 2021.

Religious transnational tourism has received some sustained scholarly.[2] Pilgrimages across long distances are an age-old practice that, as the hajj and the camino de Santiago de Compostela show, has been taking place well before the rise of nation-states with the Peace of Westphalia in the 17th century. In contrast, trans-national religious tourism, that is, religious tourism that involves the building of multiple, simultaneous, and constant relationships across national borders, is much newer and has received relatively little attention. The "East," which in the Western imagination is commonly portrayed as more "spiritual" and "mystical," is often the destination of these trips.[3] The combination of religion and tourism has become one of the most significant transformations in the Chinese religious landscape. Chinese scholar Ma Jinfu found that, in 1997, there were over 2233 sites of religious tourism in China, including 1630 Buddhist, 335 Daoist, 240 Islamic, and 18 Christian sites.[4] Among them, many famous temples and pilgrimage sites, such as Buddhist and Daoist mountains, are major tourist sites. Every year, thousands of pilgrims travel to these religious sites to fulfill their religious pursuits. In addition, increasing numbers of tourists have been attracted by these religious destinations for their beautiful natural surroundings and rich cultural and historical heritage.

The extant literature notes how the desire for authenticity and commodification, two themes that I will discuss in my article, shape and result from the interplay of religion and tourism.[5] My contribution to this scholarship is the use of the notion of transnationalism to explore how the cult of Mazu—more specifically, the way the devotion to the goddess at her ancestral temple in Meizhou—has changed as a result of pilgrimage and religious tourism from Taiwan and by Chinese people living in other countries abroad.

The modern transformation of Mazu worship is part and parcel of significant changes in the Chinese religious arena as a result of new policies of the Chinese Communist Party (CCP) regarding freedom of religion and the preservation and promotion of cultural heritage. These changes have been combined with the deepening impact of globalization on China's economy and culture. One of the most visible examples of this combination is the dramatic growth of religious tourism, both within China and from the diaspora. Starting from the end of the 20th century, Chinese religious traditions, including Buddhism, Daoism, and other popular religions, have seen many of their ancient sacred spaces become increasingly commercialized and folklorized into tourist sites. According to Juan, around 130,000 religious sites in China have been turned into tourist attractions.[6] The case of Meizhou temple provides a good window into the development of religious tourism in China, allowing us to see the incentives and characteristics of this phenomenon and highlighting differences between traditional religious activities and modern innovation.

Today, the Meizhou ancestral temple is one of the most important tourist sites that blends long-standing popular religious beliefs and rituals with cultural pilgrimage and tourism. This is not surprising, given that Mazu is the most prominent Chinese goddess, and that her cult has been transmitted not only all over China, but also to over 20 different countries, following the steps of Chinese diaspora. In fact, while the devotion to Mazu is particularly strong in southeast Asia, Mazu has become a pan-Asian object of worship, transmitted to all throughout of the world. As Lee's work demonstrates, the term transnationalism implies more than just simply crossing national borders sporadically. Here, I adopt the seminal definition of Linda Basch: "transnationalism is a process that involves the forging and maintenance of "multi-stranded social relations that link together [the migrant's, tourist's, or traveler's] societies of origin and settlement".[7] We shall see how the worship of Mazu creates shared social spaces and builds networks in which and through

[2] See Norman (2013), Raj and Griffin (2015), Rocha (2006), Stausberg (2011), and Timothy and Olsen (2006)
[3] Lanquar (2011).
[4] See (Tan 2015, p. 27).
[5] For example, see Bremmer (2020).
[6] Juan (2010)
[7] (Basch et al. 1994, p. 7)

which the Chinese in diaspora experience close and strongly affective connections with mainland China.

Contemporary Chinese popular religious practices have received some scholarly attention. For example, Chau Adam Yuet, John Lagerwey, and David Johnson have studied the revival of Chinese popular religious practices such as temple festivals in rural China.[8] All of these scholars understand this revival to be a reflection of or to go hand-in-hand with contemporary economic and political developments, such as the influence of market economy and the state's new religious policy in the 1990s that opened up more spaces for worship. However, they did not pay enough attention to the role of religious tourism in the revival of Chinese popular religions.

In his study on the Mazu culture derivative and products, Liao Xhien Chih explores the role of tourism in Mazu's cult, showing how temples to the goddess in Taiwan have adapted some traditional styles of Mazu's images to design new artifacts to attract more pilgrims and appeal, in particular, to younger generations.[9] Wang Qingsheng also discusses how the interaction of culture, local industry, and tourism led to the development of the Dajia Mazu International Tourism Culture Festival in Taiwan,[10] Wang details a series of policies applied by Taiwan governments to promote traditional culture and tourism. Moreover, he explores the special role that performances by art troupes have had in the development of the festival, which has established the Dajia temple as the inheritor of traditional culture. Huang Ying-Fa's article "From Religious Pilgrimage to Tourism and Bodily Cultivation: Taiwan Dajia Mazu Pilgrimage" also focuses on Taiwan Dajia Mazu and its transformation from religious pilgrimage to cultural tourism.[11] works provide insightful perspectives on religious tourism in the Taiwan area. Recognizing the socio-political and cultural conditions in mainland China, I would like to complement them by offering a view from Mazu's ancestral temple.

This article builds upon but goes beyond these works, by focusing on transformations linked to tourism in the ancestral temple of Mazu in Meizhou Island, the sanctioned point of origin of the traditions and cult around the goddess. The key questions I pursue are: how has the ancestral temple transformed itself from a place of ancient religious traditions publicly associated with provincial superstition into a famous pilgrimage and tourist site that represents Chinese cultural heritage and serves as a source of identity and belonging for the Chinese diaspora? How has Mazu's image changed as a result of these transformations? To answer these questions, I examine three major transformations in the devotion to Mazu at the ancestral temple at Meizhou. First, architecturally, the ancestral temple has been refurbished into a national tourist resort, which provides a full array of recreational and leisure services to pilgrims and tourists, in addition to performing traditional rituals. Second, these rituals are no longer just a means to connect devotees with the goddess, but are now performed as a foundational part of China's cultural heritage. Third, the Meizhou temple reemphasizes its religious hegemony as the original site of Mazu worship by combining traditional rituals such as "dividing incense" with new practices like "Mazu, returning to mother's home from all over the world." In the conclusion, I argue that the new developments of Meizhou temple illustrate the political and economic context within which contemporary popular religious traditions are reconstituting themselves in order to survive. These developments also reveal the interaction of popular religious traditions with state interests and economic dynamics, especially modern tourism.

8 Chau focuses on the social aspects of popular religion in modern North and North-central China. Through a case study of Longwang gong in a village of the Shanbei area, he illustrates the role the local government played in the revival of popular religion Chau (2006). Johnson pays special attention to rituals and temple festivals in villages in Shanxi province Johnson (2009). Lagerwey provides detailed descriptions of popular Chinese festivals, including pan-Chinese festivals and local festivals dedicated to local deities Lagerwey (2010).

9 Liao (2019).

10 Wang (2018).

11 Hung (2009).

2. The Transformation of a Local Religious Temple into a National Tourist Resort

This section will focus on the way that the Meizhou ancestral temple has evolved from a traditional temple complex to a tourist resort that attracts millions of tourists annually. We will see that the temple association, which was established in the Song dynasty to oversee the site, has played a key role in this evolution through the preservation of traditional architecture and religious–cultural artifacts of historical significance and through the building of new structures more in line with the demands of a growing tourist industry. I will have more to say about the structure and transformation of the temple association and the management of the temple in response to tourism later. For now, I would like to historically trace the architectural evolution of the temple from its origins to the present.

The earliest account of the Meizhou ancestral temple appears in *Shengdun zumiao chongjian shunji miaoji*, compiled by Liao Pengfei in 1150, during the Song dynasty.[12] According to Liao,

> It was said that she was the goddess was capable of communicating with Heaven. Her surname was Lin and she lived at Meizhou Island. At first, she lived as a female shaman and had the ability to forecast fortune and misfortune. After she died, people enshrined her at this island.[13]

As we can see from this historical record, the first temple to worship Mazu was a small shrine established by local people. Historical sources show the expansion and development of this simple shrine into a full-fledged ancestral temple. A mythical story in "*The Record of the Founding Temple in Meizhou Island*"[14] sheds light into how this process took place during the Song dynasty:

A merchant, Sanbao, was on his business trip to overseas. Due to bad weather, his ship had to anchor in the Harbor of Meizhou Island. When they were about to leave the next day, the anchor of the ship became stuck on a heavy stone, which could not be removed. After acknowledging that the goddess's shrine was extremely efficacious, the merchant made a pilgrimage to the shrine, offered incense to the goddess, while praying to the goddess, "If we can return safely home through storms, I will donate a large amount of money to establish a temple for you to thank for your numinous assistance". With the protection of the goddess, the merchant returned home safely. When the ship was passing through the island, the merchant made his pilgrimage to the goddess's shrine again. To fulfill his vows, the merchant donated money to enlarge the shrine into a temple in Meizhou.[15]

The enlargement of the temple complex in Meizhou during the Yuan dynasty stemmed from Mazu's divine protection of rice transportation. In the second year of the Tianli era (around 1329), the official who was responsible for rice transportation from the capital to Sansha city experienced a seven-day storm at sea. After crying out to the goddess for help, divine lights appeared, and the fleet was guided safely through the storm. Subsequently, the government official donated money to refurbish the ancestral temple[16]. In this process of establishment of Mazu's ancestral temple, we can see the on-going role of the state in sponsoring it and in the promotion of the Mazu cult, in recognition of her divine involvement

12 (Li 1995, p. 2).

13 The original text comes from Liao Pengfei, *Shengdun zumiao chongjian shunji miaoji*, collected in (Jiang and Zheng 2007, pp. 1–2).In the process of establishing a canonical, glorious history of the Meizhou temple, alternative narratives of Mazu's origin were erased or ignored. According to (Watson 1985, p. 297), some Taiwan villagers believe that Mazu emerged as the daughters of a poor fisherman. People in the New Territories think that Mazu was the daughter of danmin (an ethnic group of people who live on the boat, referred to as "Tanka" in Cantonese). These origin stories, which dispute the official version constructed around the Meizhou temple only circulate as local or regional oral accounts. Imperial patronage was key in establishing the Meizhou temple's version as foundational and canonical.

14 (Jiang and Zheng 2007, p. 80).

15 (Jiang and Zhu 2011, p. 90).

16 Ibid., p. 91.

to protect it and advance its interests. As I have discussed elsewhere, state patronage of the construction of Mazu temples was an important component of her canonization[17].

While all these initiatives consolidated the shrine at Meizhou as a visible and ritually efficacious site, it is only in the Ming and Qing dynasties that Mazu worship reached its heyday. By then, the temple had become a large complex, which included "the Palace of Celestial Empress" (*tianhou dian*), a "Facing Heaven" garret (*chaotian ge*), a drum and bell tower (*zhonggu lou*), a dressing tower (*shuzhuang ge*), the temple gate, and Taizi Palace (*taizi dian*). Each building of the temple was glorified by a divine manifestation of the goddess as shown in the *Record of the Sagely Manifestation of the Celestial Consort* (*Tianfei xiansheng lu*).[18]

"Pushing over Waves to Help Ships Cross a Storm," (*Yonglang jizhou*), describes the reconstruction project launched by the Commander Zhou to honor the goddess's divine protection in assisting Ming military ships to fight pirates:

> In the seventh year of the Hongwu reign (1374), the commander of Quan Prefecture, Zhou Zuo, [led] warships to patrol and arrest [pirates]. Suddenly, they encountered a strong hurricane springing up. The anchor was broken up and the ship ran aground. Sailors on the ship cried from all sides, knocked their heads, and cried for the goddess's help. Shortly later, there was a divine light appearing in dark night which suspended in mid-air to illuminate everything together. The divine light also appeared upon the masthead of the ship.[19]

The divine light was believed to be a sign of Mazu's divine protection. Mazu was able to push over huge waves to send military ships to the harbor safely. This story ends with Commander Zhou's devotional activities to sponsor the reconstruction of the Meizhou ancestral temple, including building the incense pavilion, drum and bell tower, and the temple gate.[20]

The following episode provides accounts of the establishment of the Facing Heaven Pavilion: "As Commander Zhang conducted warships to patrol the sea, he prayed for the goddess's blessing. The goddess responded to his prayer, as expected. Commander Zhang then shipped building materials to the Meizhou Island and built a garret on the left side of the central hall, named 'Facing Heaven Garret'".[21] Once again, we see here that the expansion of the temple was due to the patronage of mobile groups, whether merchants or imperial officials. In this case, it is a military man charged with projecting the empire's naval power. This is why we can affirm that the expansion of ancestral temple is part and parcel of the state's deployment of Mazu as a metaphor for imperial power.

Similar narratives, in which devotees donate money to enlarge the temple, appear in different contexts. That is especially the case in "The Grand Governor Writing Prayer and Memorial" (*zongdu zhudao shuwen*) and "Building Dressing Tower in Gratitude" (*zhuanglou xieguo*). These two episodes contain several common elements. First, the goddess manifested herself to provide divine protection to devotees who prayed to her. Second, the devotees donated money or building materials to reestablish the temple in honor of the goddess' numinous responses. While the Grand Governor Yao Qisheng (1623 to 1683) established the temple building named Taizi Palace, a local devotee sponsored the establishment of the dressing tower.

The above review of the history of Mazu's ancestral temple at Meizhou illustrates the way that the temple evolved from a small shrine into a splendid complex. That each

17 For a detailed study of the role of imperial governments in the origin and development of Mazu cult in late imperial China, see (Zhang 2019). Building on Stephan Feuchtwang's notion of "imperial metaphor," I argue that the state was central in the canonization of Mazu, elevating her from a polyvalent local object of devotion to a key deity in a national cult that strongly reflected official ideology. State canonization involved three deeply intertwined strategies to standardize, give public recognition, and promote a range of local beliefs and practices: the conferral of official titles, the incorporation of local gods and goddesses into the register of sacrifices, and the construction of official temples.

18 *Tianfei xiansheng lu* (*Record of the Sagely Manifestation of the Celestial Consort*), compiled by Zhaocheng (c. 1644). The text used in this paper comes from (Jiang and Zhou 2009, pp. 68–103).

19 *Tianfei xiansheng lu*, pp. 96–97.

20 Ibid.

21 *Tianfei xiansheng lu*, pp. 96–97.

of the additions and enhancements to the original shrine was associated with a miracle performed by Mazu serves to legitimize and glorify the temple, while simultaneously crafting mythical narratives about the salvific power of the goddess. In other words, the growing size and elaborateness of the building go hand-in-hand with her rising status as an efficacious sacred figure. This tight connection between the glorious history of Mazu's divine interventions on behalf of her economic and political influential devotees and the gradual-yet significant enlargement of the temple have contributed to making the Meizhou ancestral temple the preeminent sacred site for the worship of the goddess.

The history of the ancestral temple is, however, not a linear one of increasing prominence. During the Cultural Revolution in 1966, Mao launched an attack on all traditional ideas and things, including all religious institutions, activities, and objects. Among other measures, he ordered the closing of all religious institutions, the laicization or imprisonment of clergy, and prohibition of public and private religious activities. In the case of the Meizhou ancestral temple, most of its buildings established in the Qing dynasty were destroyed, as were the three statues of Mazu also constructed during this dynasty.[22]

Following the death of Mao in 1976, his perpetual revolution began to ebb. Taking advantage of the new context, the temple association, along with local devotees of Meizhou Island, initiated the project to renovate the ancestral temple complex in 1978.[23] Through this renovation project, the temple complex has preserved the traditional architecture style of the Qing dynasty. This process of remodeling the traditional architecture style can be seen as a case of the "invention of tradition," which Hobsbawm and Ranger characterized as "a set of practices, normally governed by overtly or tacitly accepted rules and of a ritual or symbolic nature, which seek to inculcate certain values and norms of behavior by repetition, which automatically implies continuity with the past"[24]. The "re-creation" of the traditional buildings establishes an unbroken link with Mazu's place of birth, further legitimizing the temple as the undisputed inheritor of ancient religious tradition.

From 1998 to 2002, the temple association embarked on a major enlargement project to meet the demands of increasing numbers of pilgrims and tourists.[25] A new wing was constructed, consisting of new modern style structures, such as the square of the Celestial Empress, an opera stage, a temple gate, a drum and bell tower (*zhonggu lou*), the Palace of Timely Salvation (*shunji dian*), the Palace of the Celestial Empress (*tianhou dian*), the Palace of Praying for Blessing (*qifu dian*), an exhibition hall of Mazu culture, the Hotel of Praying for Blessing (*qifu binguan*), a stone statue of Mazu, and the group sculpture of Mazu's story.

As we can see from the list of new Mazu temple buildings, the temple association built several structures in line with the development of tourism. For example, the square of the Celestial Empress, which covers 10,000 square meters, sided with viewing stands to accommodate almost 10,000 tourists, is designed to perform large sacrificial ceremonies and other recreational activities. This massive space stands in contrast to the smallness and intimacy of the early shrine. The Palace of Praying for Blessing serves as a place where pilgrims, devotees, and tourists can invite Mazu statues to their homes and light lanterns of blessings. The exhibition hall of Mazu culture showcases her relics, antiques, paintings, and calligraphies. The Hotel of Praying for Blessing provides expanded accommodations for tourists. Before the new constructions of the hotel and other structures, the pilgrimage of devotees from throughout the world posed an enormous logistical problem for the Meizhou temple, as it could not accommodate all of them during the ritual services and performances. A saying circulated among the locals: tourists to Meizhou Island "visit the temple at daytime, while they sleep at night." Indeed, one pilgrim from the Taiwan area I interviewed told me that he really enjoyed staying in the hotel at night, visiting the temple, and sightseeing the beautiful environment of the Meizhou Island during the daytime.

22 See (Jiang and Zhu 2011, p. 94).
23 See (Jiang and Zhu 2011, p. 94).
24 See (Hobsbawm and Ranger 1983, p. 1).
25 See (Jiang and Zhu 2011, pp. 100–3).

A 14.34-m-high stone statue of Mazu entitled "the Goddess of Peace" is one of most popular spots for tourists to take pictures and enjoy the scenic view (Figure 1). It is part of a group of statues that render in modern form Mazu's story and myths as recorded in *Pictorial Record of Sagely Manifestation from Holy Mother of Celestial Empress* (*Tianhou shengmu shengji tuzhi*).[26] We can conclude, thus, that tourism is shifting the emphasis in the devotees' relation to Mazu from the textual representations that circulated widely in pre-modern China and were central to the literati's construction of her myths to large-scale material representations such as buildings, plazas, and monuments that provide opportunities for tourists to have enjoyable and easily recordable experiences. In addition, the cult of Mazu is increasingly becoming a blend of traditional religious practices and non-religious aesthetic experiences. For instance, the temple association has built some scenic spots to emphasize the beauty of the island's landscape, such as the pavilion of sea view, the pavilion of sunrising, the pavilion of listening to the roaring water. These scenic spots allow visitors to experience the sacredness of the place beyond the confines of the temple.

The incorporation of the natural environment through place-making practices of landscaping has been critical to transforming the temple into a national tourist resort (See Figure 2).[27] According to temple history, the Meizhou ancestral temple was under the charge of religious specialists, starting with the monk Shi Zhaocheng (c. 1644) in the late Ming dynasty. The successive generations of temple chairs were inherited from masters to disciples. This genealogy of temple chairs has been abandoned in modern China. To adapt to the changing nature of pilgrimage and religious tourism, the Meizhou ancestral temple has evolved from a religious association managed solely by religious specialists to an entity that integrates religious, economic, and tourist services. Starting from 1986, the temple has been managed by a temple board, consisting of a chairman and board members.[28] The temple board oversees and manages departments created to improve the temple's ritual and tourist services. There is now an administrative office for scenic attractions. Taifu company, the temple-owned enterprise, which is affiliated to Antai Hotel and Qifu Hotel, has been contracted to provide accommodations for tourists. In addition, an independent labor service company is responsible for producing the tourist merchandise, as well as managing stores, booth rentals for the temple attraction areas, and Mazu's pastry shop. The art troupe, Celestial Empress, performs traditional music, dancing, and local opera to attract tourists, while a chanting troupe offers chanting services for Mazu devotees and an "offering production group" is charged with making sacrifice rituals to the goddess. Finally, a medical clinic dispenses medical care for tourists and pilgrims. In other words, tourism has led to the proliferation and formalization of temple functions, as well as to the commodification of religious practices and artifacts.[29] This commodification does not mean that religion is simply a reflection of economic forces. Rather, there is a relation of reciprocal determination between religion and economy. Mayfair Yang argues that the temple's "ritual economy cannot be seen merely as the result of economic development, for ritual life has also fueled economic growth (it often provides the organizational apparatus, sites, and motivation for economic activity).[30]

The local government has played a major supporting role in the process of architectural, spatial, and managerial transformation of the ancestral temple. This is because the temple's high economic and tourist potentials are considered effective vehicles for generating prosperity for the local economy and income for local government.[31] Since 1992, China's central Communist government has designated Meizhou Island as a national tourist resort, further increasing the profile of the ancestral temple. To develop the tourism to Meizhou, the Putian government sponsored the temple association to refurbish the

26 *Tianhou shengmu shengji tuzhi* contains a set of hagiographic paintings depicting the miracles performed by Mazu. It was compiled around 1826.
27 For more on religious place making, see Vasquez and Kim (2014).
28 (Jiang and Zhu 2011, pp. 107–9).
29 Ibid., pp. 109–10.
30 (Yang 2000, p. 480).
31 See (Poceski 2009, p. 255).

buildings and religious sites of the ancestral temple. As described in the local government document entitled "Construction Planning of Ancestral Temple", the reconstruction project was envisioned to "highlight the traditional style and thereby making the temple complex as an ideal of ancient architecture. This temple should be a combination of religious pilgrimage, tourism, academic center, and vocation".[32] In other words, key to the invention of tradition was making the ancestral temple a "foundational hierophany," to draw from Mircea Eliade, an axis mundi connecting to primordial, mythical times.[33]

Figure 1. The statue named "the Goddess of Peace." Source: Author's photograph.

The invention of tradition, commodification, and architectural reconfiguration of the ancestral temple in Meizhou has made it a major tourist and pilgrimage destination that attracts millions of tourists, including Mazu devotees from oversea. According to the Putian government, the Meizhou temple accommodated around 1.56 million visitors from both mainland and overseas China in 2010.[34] These devotees bring the capital that they have earned through their hard work and success in diaspora, further encouraging more development and commodification. On the one hand, the promotion of Mazu temple by the local government in the Meizhou area has contributed to the transformation of the Mazu cult; on the other hand, this promotion has brought financial income to support local economic and social developments.

32 See (Liu 2011, p. 253).

33 Eliade (1959).

34 See http://www.putian.gov.cn/zjpt/qhgk/xzqh/201210/t20121031_125384.htm, accessed on 17 February 2021.

Figure 2. The map of Meizhou Island and the Mazu Cultural Tourist Site with various religious, cultural-historical, and entertainment attractions. Source: Author's hand-drawn map.

3. Reproducing Traditional Culture and Inheriting Legacy: Folklorization (*minsu hua*)

The construction of the translocal devotion to Mazu was part of a gradual process, in which politics, that is, the actions of and recognition by local and imperial governments, have played a major role, setting the stage for the contemporary, tourist-based promotion of the cult. Let us review this process. According to tradition and official historical records, Mazu worship was incorporated into the Register of Sacrifices (*sidian*) starting from the Song dynasty.[35] In the Song era, the Register of Sacrifices referred to a list of ceremonies performed by the emperor and his officials, including rituals performed by local officials at local shrines. Preparing and participating in ritual performances was one of the main responsibilities of local officials. Since then, the Mazu cult was incorporated into the official pantheon under the administration of the Imperial Board of Rites, which standardized sacrifices dedicated to Mazu in accordance with the regulations of the Register of Sacrifices starting from the mid-11th century. As recorded in *Songhui yao*, prefects and magistrates were obligated to perform official ceremonies at local Mazu temples recognized by the Song government twice each year, in the mid-months of spring and autumn. In addition, the emperor also dispatched officials to perform official sacrifices in the ancestral temple of Mazu at Meizhou as ordered by imperial edict. The performance of the official sacrifice ceremony ordered by imperial edict was considered as the greatest honor for local cults.

The Qing government established a more routinized and standardized approach to the official rituals dedicated to Mazu. She was incorporated into the official sacrifice system, as can be seen in the regulations presented in the Qing's Register of Sacrifices from 1733. Since then, the highest-ranking bureaucrat in every part of the country—each province, prefecture, and county—had the duty of worshipping Mazu during the spring and the autumn. In addition to the spring and autumn sacrifices, Qing emperors also issued edicts to dispatch officials from the central government to deliver sacrifices at the ancestral temple

[35] The practice of keeping a register of sacrifices can be traced back to the Book of Rites. In imperial China, including the Tang dynasty, the register of sacrifices primarily recorded the sacrificial rituals performed by the imperial government. It also contained detailed information regarding the categories of offerings and the sites for performing rituals.

of Mazu at Meizhou Island on behalf of the emperors. Qing official texts, such as "Collected Statutes of the Great Qing from the Kangxi Reign" (*Kangxi daqing huidian*), stipulated the proper rules, procedures, and materials to be used in official rituals, including the arrangement of orchestras, the literary format of eulogies, ritual garments, honor guards, and the lists of officials who could witness the events.[36] The dates of the standard sacrifices and the offerings first had to be approved by the emperor. The dispatched official and the local officials obligated to perform the ceremony were representatives of the emperor, and thus they ritually enacted the emperor's attitude of reverence for the deity. In addition, the official prayers written for the sacrifices devoted to the deities elaborately illustrate the emperor's authority. Specifically, the prayers make it clear that the emperor sent officials to Mazu temples to announce the imperial title of the goddess granted by the court and to reward her contributions to the imperial government.

To give a sense of the ritual etiquette involved, the following is a sketch of the official rituals dedicated to Mazu. The preparations involved the participants purifying themselves for two days. Everything had to be put in order on the day of the ritual, including the three sacrificial animals that would be offered and the many sacrificial instruments used at the occasion. Members of the music office (*jiaofang si*) played the music. The cantors (*zanyin guan*) would lead the sacrificer (*chengji guan*) up to the left gate and into the dressing room. After purifying themselves, the officials assumed their assigned positions in the temple hall. The ceremony began with the ritualist "welcoming the deity" (*yingshen*). The sacrificer and his assistant presented incense to the deity in front of the altar three times, and performed a ritual sequence of bowing, prostrating, and rising. The ritualist then announced that he would "proceed with the first sacrifice." Silk, libations, and prayers were offered on the altar by the officials. The master of prayer read the written prayers, placed them on the altar, prostrated thrice, and withdrew. The second and final offerings included two libations: the master of wine offered the wine vessel at both the left and right sides of the altar, and then returned to his place. The ceremony ended by "bidding farewell to the deity" (*songshen*). All the officials performed three prostrations and nine kowtows, and the prayers, silk, and food were sent away to be burned.

These descriptions of the intricate ritual etiquette at Mazu's ancestral temple show that it was mostly a top-down affair that required extensive training. It is also clear that the temple had neither the facilities nor the activities to create a total immersive experience for pilgrims and tourists alike. The temple board had not only to transform the sacred space, as we saw above, but also had to enhance its attraction and reputation by renewing its traditions. With this aim in mind, the temple board sought to retain the "original," official sacrifice ceremony, while staging it for a large audience. At the first tourist festival of Mazu culture in 1993, the performance of this reconstructed official sacrifice ceremony became the centerpiece. Since then, the official sacrifice to Mazu at Meizhou, along with the Yellow Emperor sacrifice at Shan'xi and the Confucius sacrifice at Qufu, have been designated as national sacrifices by the Ministry of Culture.

Despite its claim to be fully grounded in ancient traditions, the restored official sacrifice of Mazu integrates into the traditional ceremony some modern innovations to attract tourists, meet the needs of devotees, and express the current times. In terms of following ritual traditions as described in the official sacrifice of the Qing dynasty, the modern version incorporates the basic traditional ritual structure, including welcoming the goddess, offering incense, offering silk, three sacrifices, the performance of three prostrations and nine kowtows. To show the continuities and variations vis-à-vis the ritual etiquette in the Qing version, here is a sketch of the modern version dedicated to Mazu (I participated in the sacrifice ceremony dedicated to Mazu held at the 21st China Meizhou Mazu Cultural Tourist Festival on 1 November 2019. The description of this

[36] The original text comes from *Kangxi daqing huidian*, collected in (Jiang and Zhou 2009, pp. 197–99).

sacrifice ceremony is based on my fieldwork observation and the record of Mazu temple at Meizhou Island).[37]

"The banquet of sacrificial offerings," a long dining table, is placed on the platform of the Central Hall. Mazu's sedan and incense altar are situated in the central position of the dining table, symbolizing the goddess's status and presence at the ceremony (Figure 3). The area in front of the altar is the area to make offerings to Mazu and to read the eulogy honoring the goddess. The right sides of the altar are the place for purifying. The stage below the banquet contains the leading and assistant sacrificers. The dining table displays sacrificial artifacts and offerings. In line with the traditions, the sacrificial offerings adopt animal offerings of *shaolao* (lesser lot), which includes a pig and a sheep.[38] The offering items for three sacrifices include wine vessels, fruits, and flowers.

Figure 3. The leading and assistant sacrificers on the platform of the Central Hall. Source: the website of Meizhou ancestral temple https://weibo.com/mzmz323?reFigure1005055013.is_all=1#_rnd1610641535373. Accessed on 10 July 2020. Reprinted with permission from the temple board of Meizhou ancestral temple.

First, the cantor announces to the fire the salute and beats the big drums as the ceremony opening. The cantor then proclaims that the guard of honor and guardian will assume their position. The leading and assistant sacrificers are led by the ritualists to assume their proper positions (Figure 3). The performance group, including dancers, musicians, and singers, also assume their proper positions. After that, the songs to welcome the goddess are played. The leading sacrificers cleanse their hands, pick up the incenses, and present incense to Mazu in front of the altar and bow three times. The ritualist will collect the incense and put it in the incense burner. The cantor announces sacrificers will offer the sacrifice of silk. The leading sacrificer will read the eulogy in the traditional style. The ceremony proceeds with the first sacrifice accompanied by playing the music of "the Peace of Ocean." The wine vessels are offered on the altar, followed by a second sacrifice accompanied by the music of "Peace." Fruits are offered; the last sacrifice is performed with the music of "Universal Peace" and the silks are presented to the goddess. During the ceremony of three sacrifices, the dancers will perform the traditional dance, *Bayi*, with

37 I participated the sacrifice ceremony dedicated to Mazu held at the 21st China Meizhou Mazu Cultural Tourist Festival on 1 November 2019. The description of this sacrifice ceremony is based on my fieldwork observation and the record of Mazu temple at Meizhou Island. See (Jiang and Zhu 2011, pp. 370–71).

38 The animal offerings of *shaolao* including a sheep and a pig, is present in the area in front of the altar before the sacrificial ceremony.

plumed artifacts. The ceremony ends by burning the prayer scrolls and silks, and "bidding farewell to the goddess." All the sacrificers perform three prostrations and nine kowtows.

The modern version of sacrifice ceremony bears strong family resemblances and shares a common temporal template with the traditional version. As historian Michel-Rolph Trouillot has argued, the selective management of time and memory, what is preserved and what is not, is one of the ways by which elites reproduce power.[39] These two versions of sacrifice ceremony are both organized around the sequence of preparing the ceremony, greeting the goddess, presenting the incenses, reading the prayer, performing three sacrifices, and sending off the goddess. In addition, there are other traditional elements reflected in the modern version, such as the sacrificial offerings, artifacts, musical instruments, and the outfits of cantors and ritualists. The application of traditional elements is to emphasize the glorious history of this sacrificial ceremony. In this sense, the modern version performed by the Meizhou temple is promoted as an expression of the ritual heritage of imperial China and further confirms its religious hegemony as the first Mazu temple in the world.

Nevertheless, to attract modern tourists, there are some innovations in the event. First, the sacrifice ceremony of Meizhou temple adds the performance of *yuewu* (dance with music) during the section of three sacrifices. This traditional dance, which is accompanied by sacrificial music, *yuewu*, represents an important innovation on the traditional sacrificial dance, *bayi*, which is normally used in the sacrificial ceremony dedicated to Confucius and performed by a male dancer. To emphasize the Mazu's identity as goddess, *yuewu* are only performed by 20 year-old female dancers whose body postures are meant to express the female image of a compassionate sea goddess: "a beautiful young girl who saves people in the sea".[40] (Figures 4 and 5). The dance also uses plumed artifacts and combines some postures from local opera troupes to highlight Mazu's femininity and her local background. The music that accompanies *Yuewu* is composed by modern musicians Lin Hanzu and Zheng Ruilin, incorporating the music of "welcoming the goddess," and the music of "Peace of Ocean," "Peace," and "Universal Peace".[41] This sacrificial music inherits the style of traditional sacrificial music to emphasize the solemnity and holiness of the ceremony, while incorporating melodies from local music, adding a folkloric element to the performance. Second, the cantor and ritualists are trained female actors who are college students in a local university (Figure 6). In the traditional sacrificial ceremony, the cantor and ritualists are male officers from the Ministry of Rites. According to the ritual history, women were not allowed to participate in official ritual ceremony. To create an entertaining and alluring atmosphere, the modern version includes female actors with beautiful dancing postures and a female cantor with a nice voice who functions as a host (Figure 5).

Third, the leading sacrificers are usually the chair of the temple board from the ancestral and affiliated temples. In the sacrificial ceremony I observed, the leading sacrificers were the chair of the temple board of the ancestral temple at Meizhou Island, Lin Jinzan; the temple chair from the Celestial Empress Temple at Taiwan Beigang (*beigang tianhou gong*), Cai Yongde; the temple chair from the Celestial Empress Temple and the Selangor and Federal Territory Hainan Association in Malaysia, Deng Cairong. In other words, the religious specialists reflect the transnational character of the devotion, including ritualists from key nodes in the diaspora, nodes that are marked by prominent-yet-secondary temples. In the same way, the assistant sacrificers consist of members of the temple board and overseas pilgrims. The move to invite temple chairs and pilgrims from abroad is to reinforce the relationship between the ancestral temple and its affiliated temples beyond the mainland and to generate a unified sense of identity and belonging among all pilgrims who are main customers and donors for the temple attractions at Meizhou. It represents an effort to extend the networks of patronage that were established during the imperial

39 See Trouillot (1995). In the process of repacking the traditional ritual of sacrifical ceremony, we should acknowlege that other local and regional histories of Mazu has been erased as the cult has been canonized and anchored in Meizhou.

40 See (Jiang and Zhu 2011, p. 368).

41 Ibid., p. 367.

period. As I described above, in the traditional sacrificial ceremony, the leading sacrificers were officers dispatched by the emperor, while the assistant sacrificers were local officers. In this traditional version, the dispatched official and the local officials charged with performing the ceremony were representatives of the emperor, and thus, they ritually enacted the emperor's attitude of reverence for the deity. In contrast, in the modern ceremony, the reverence that is enacted and that has become central to the religious legitimacy and visibility of the cult is that of the pilgrims and tourists, who are increasingly coming not only from throughout China but also from the Chinese diaspora.

Figure 4. Female dancers performing *yuewu*. Source: the website of Meizhou ancestral temple https://weibo.com/mzmz323?refer_flag=1005055013_&is_all=1#_rnd1610641535373. Accessed on 10 July 2020. Reprinted with permission from the temple board of Meizhou ancestral temple.

Fourth, the sacrificial ceremony is held on Mazu's birthday, the day of "her ascending to Heaven", and of Meizhou's Mazu Cultural Tourist Festival, a coordination that seeks to increase its tourist appeal. In addition, the performance teams are divided into small-, medium-, and large-scale groups to meet the different ritual requirements. For example, the small-scale group is responsible for providing sacrificial ritual services for pilgrimage groups. This group also performs an abbreviated version of the sacrifice ceremony, which lasts just 12 min, every Sunday morning, making it possible for tourists with busy itineraries to attend, while also requiring them to stay overnight to be able to make it to the ceremony. In turn, the large-scale team performs the traditional sacrifice ritual at the ancestral temple's Mazu square, which, as stated above, covers 10,000 square meters and is sided with viewing stands to accommodate almost 10,000 tourists.

In the process of restoring the sacrificial ceremony, the temple association appropriates old ritual traditions, adapting them and blending them with new practices and media to respond to the demands of tourists and to capture the financial benefits brought by tourism and government agencies. This is a process of "folklorization," in which local religious and cultural traditions are reinterpreted as "part of the folklore," expressions of the core cultural identity of a people. This identity can then be preserved, performed, consumed, and marketed nationally and globally as an authentic, intangible cultural heritage. As Zhou argues, cultural "heritagization" becomes a new means of seeking legitimacy for popular religious practices: "Before the rise of intangible cultural heritage, components of folk beliefs such as shrines and temples were protected due to their value as 'antiquities.' Now that these places have been declared by the government as 'heritage conservation units,' the folk beliefs associated with them have indirectly legitimized".[42] What is underlying the processes of folklorization and cultural heritagization is a powerful desire for authenticity, a desire by the pilgrims and tourists alike to experience tradition as its most essential, as it was at its origins[43].

Figure 5. The performance of *bayi*. Source: the website of Meizhou ancestral temple https://weibo.com/mzmz323?refer_flag=1005055013_&is_all=1#_rnd1610641535373. Accessed on 10 July 2020. Reprinted with permission from the temple board of Meizhou ancestral temple.

Figure 6. Female actors who play ritualists and cantors. Source: the website of Meizhou ancestral Table 323. https://weibo.com/mzmz323?refer_flag=1005055013_&is_all=1&ssl_rnd=1616329203.3123#_rnd1610641535373. Accessed on 10 July 2020. Reprinted with permission from the temple board of Meizhou ancestral temple.

42 Ibid., 158.
43 See Bremmer (2020).

In the case of the Meizhou ancestral temple, the sacrificial ceremony at the temple has become the national model for recovering and transmitting traditional culture. Moreover, the United Nations' Educational, Scientific, and Cultural Organization (UNESCO) identified the temple and its reconstituted ceremonies as part of humanity's intangible cultural heritage in 2009. As a result, the temple now benefits from special protection and national funding and can thus be repackaged as something that has universal legal standing and recognition. In effect, this is a new national and global "canonization" of Mazu.

Now, the goddess is not an imperial metaphor, but a trope for promoting religious traditions outside the realm of religion, for circulating traditions stripped of their local, grassroots signification and presented as living inheritors and protectors of Chinese culture and history. This resignification makes them worthy of preservation and national and internal support for politico-economic purposes, including the development of tourism.

4. Goddess of Transnationalism: Reunifying Oversea Chinese and Establishing a Common Cultural Identity

As the uncontested origin of the Mazu cult, the Meizhou ancestral temple draws from its historical prestige through a variety of strategies in order to present itself as a forge of a common cultural identity that reunifies oversea Chinese people with the motherland. A particular case in point is the traditional ritual of "dividing incense," which is combined with an innovative ritual called "returning to mother's home" (*huinaingjia*). The latter enables the temple to package itself as the "mother's home for Mazu all over the world." Furthermore, the temple also organizes other celebrations and events that consolidate its reputation as a major center for cultural, tourist, and commercial exchange across the Taiwan strait.

Mazu's popularity is highest in southeastern coastal China and Taiwan, where her temples are numerous, serving as the springboard for the cult across southeast Asia. The popularity of Mazu worship in Taiwan area was partly attributed to Qing government's military activities in the conquest of the island, to which the goddess helped with many miraculous interventions. In this sense, Mazu's presence in Taiwan served to promote the imperial ideology. After taking over Taiwan, the Qing government sponsored the construction of new temples in Taiwan's administrative centers, as recorded in the local gazetteer of the Taiwan prefecture. When the Qing government first took over the administration of Taiwan in 1727, the Intendent of the General Surveillance Circuit (*xundao*), Wu Changzuo, initiated the construction of a Mazu temple, a Guandi temple, and a Guanyin hall in the southeast area of Taiwan.[44] District magistrates in the Taiwan prefecture were obligated to sponsor the construction of official Mazu temples in their designated districts. According to Harry Lamley, during the Qing dynasty era, the construction of official temples and shrines devoted to Mazu and other state-approved deities functioned to sanction the Qing government's eventual takeover of Taiwan. He states that "This new seat of government was expected to play a civilizing role in the Ko-ma-lan region and to advance Chinese culture here".[45] To promote Chinese culture in Taiwan, the Qing officials imported images of Mazu, Guanyin, and Guandi. As the devotion to Mazu took root, it served as a mark of imperial power that inscribed the Qing government's agenda on the Taiwanese landscape and disciplined the local population.[46]

In light of the widespread popularity of Mazu in Taiwan, Tischer has argued that her devotion, particularly the collective experience of pilgrimage, allows for the creation of a "cultural intimacy," a deep sense of belonging among the Taiwanese. In turn, this personally felt, transformative ritual experience enables the production of a "spatially imagined community," that construes Taiwan as a "Mazu nation." While recognizing the role of Mazu in the construction of Taiwanese national identity, a focus on tourism calls us to adopt a different analytical scale, one that also understands the devotion of

44 "Chongxiu Taiwan fuzhi," (1760) in (Zheng 2011, pp. 210–13).

45 Lamley (1977).

46 (Watson 1985, pp. 302–3).

Mazu transnationally (Tischer himself acknowledges that Mazu is open to contestation and multiple interpretations. "To be sure, invocating a national community does not exhaust the range of interpretations borne in Mazu pilgrimages. Being claimed by the Chinese government makes Mazu a dubious ally for the Taiwanese national case, after all").[47]

In modern times, all of Mazu's devotees, regardless of their location, consider the ancestral temple at Meizhou as the sacred origin of Mazu worship and the "mother temple." Most of the "branch" Mazu temples can trace their lineage to the mother temple through the ritual tradition of "dividing the incense" (*fenxiang*) or "dividing efficacy" (*fenling*).[48] To found any new branch temple, devotees and ritualists had to go to the ancestral temple to perform this ritual. They would infuse a new statue of the goddess with the incense fragrance of the Mazu statue in ancestral temple and would then take the sacralized statue to the new temple. By sacralizing a new statue of the goddess at the ancestral temple, the new temple, in effect, drew from the goddess's efficacy and miraculous powers. The ritual of "dividing incense" also takes other forms, including picking up some incense ashes from ancestral temple's burner and placing them in an urn that is then carried to the new sacred site. By bringing these incense ashes back to the new temple building, the new temple also built a strong affinal connection with the ancestral temple. From this moment on, the new temple and its enshrined statue will have an affiliation with the older temple that puts the new temple into a subordinate position. Through this "dividing incense" system, such relations of affiliation articulate a complex network of hundreds of higher and lower temples, of "mother-daughter" temples.

This ritual tradition can be traced back to the Song dynasty. As Liu Kezhuang wrote in *Baihumiao shieryun*, "The numinous consort, a young girl, originated from the Meizhou Island through a slice of fragrant incense, through which her worship flourished all throughout Fujian province, and later spread all over China".[49] Starting from the late Ming dynasty, the Mazu ancestral temple at Meizhou began spawning offering temples in Taiwan through the ritual of "dividing incense." According to the statistics collected by Taiwanese scholars, "Since the first time an offspring temple was established in Taiwan through dividing incense, affiliated Mazu temples in Taiwan have flourished to over 2000".[50]

Since the early 1980s, Mazu's cult has served as a crucial vehicle to build contacts with the Taiwanese, attracting financial support from Taiwan and promoting the relationship between mainland and the island. According to Rubinstein, since the 1980s, the affiliated temples in Taiwan have been able to restore the bonds with the ancestral temple in Meizhou.[51] The special bond between the ancestral temple of Mazu and devotees in the Chinese diaspora attracts millions of overseas pilgrims annually. In order to sustain and renew the religious power of the Mazu temples in Taiwan, these affiliated temples periodically organize pilgrimages to the ancestral temple in Meizhou. These pilgrimages are called "incense-presenting trips." For example, Zhenlan temple in Taiwan was one of the first ones that made a pilgrimage directly to the Meizhou ancestral temple in 1987, celebrating the millennial anniversary of Mazu's ascension to heaven.[52] pilgrimage group of Zhenlan temple, consisting of 17 temple board members, brought a Mazu statue with them. When they arrived at the Meizhou temple, ritualists there performed the ritual of "dividing incense," putting the Zhenlan temple's Mazu statue in front of the altar of Meizhou Mazu to enjoy the incense and divide the goddess' religious efficacy. The sacred objects brought back with the pilgrimage group included "a Meizhou Mazu statue, a carved stone steal,

[47] Tischer himself acknowledges that Mazu is open to contestation and multiple interpretations. "To be sure, invocating a national community does not exhaust the range of interpretations borne in Mazu pilgrimages. Being claimed by the Chinese government makes Mazu a dubious ally for the Taiwanese national case, after all." See (Tischer 2018, p. 18).

[48] (Schipper 1990; Ter Haar 1990; Sangren 1993).

[49] (Jiang and Liu 2007, p. 2).

[50] Jiang (1987).

[51] Rubinstein (1994).

[52] (Yang 2004, p. 224).

an embroidered altar skirt, and the incense burner and ashes from Meizhou temple"[53] that symbolizes the renewal of the direct relationship between the Zhenlan temple and the ancestral temple. The pioneering pilgrimage journey made by Zhenlan temple set a pattern that would be repeated by other temples in Taiwan. In recent years, there has been an increasing number of religious pilgrimages by Taiwanese worshippers of Mazu to the Meizhou temple.

From the Meizhou ancestral temple's perspective, the increasing number of Taiwan religious pilgrims not only leads to more personal donations, which were crucial for the restoration of the temple in 1980s, but also to an explosion of religious tourism to Meizhou Island. Being aware of its privileged position as the most sacred pilgrimage site, the ancestral temple has developed the new ritual of "returning to mother's home" (*huinaingjia*).

This ritual is based on the tradition of "dividing incense." On special events, in particular on Mazu's birthday and the day of "her ascension to heaven," the Meizhou ancestral temple will invite Mazu devotees and affiliated Mazu temples to return to "the mother's home". The pilgrimage group, usually consisting of temple members and other devotees, bring their Mazu statue. The ancestral temple will then hold a small scale of sacrificial ceremony for the pilgrimage group. In this way, the Mazu statue brought by them also shows her homage to the original image at Meizhou. Then, the pilgrims will hold the statue of the affiliated temple and go around her with an incense burner from the Meizhou temple while whispering "Mazu is returning to mother's home." After presenting incense to the Meizhou mother, the statue of the affiliated temple will be put on the altar to enjoy the incense of the mother temple. The ritual ends with an exchange of souvenirs between temples.

"Returning to mother's home" has two meanings. First, the relationship between Meizhou's ancestral temple and other affiliated temples is understood as a mother and daughter kin relation. That is why pilgrimages from affiliated temples in Taiwan to the ancestral temple in Meizhou are seen as a "return to the mother's home." According to Steven Sangren, this kinship between the ancestral temple and affiliated temples is affinal rather than agnatic. Namely, belonging is determined through marriage, not exclusively by descent from shared male ancestors (fathers). As Sangren points out, "It is significant that the kinship metaphor used in describing the related phenomena of pilgrimages (chin-hsiang) and temple branching (fen-hsiang-literally, "dividing the incense burner") is affinal rather than agnatic. The deity images present in branch temples are similar to brides who return on a customary visit to their natal homes (lao-niangchia). During annual pilgrimages these images are taken from branch temples (or from the temples of local territorial cults where Ma Tsu maybe worshiped as a subsidiary deity) and returned to home temples where they are passed over the incense burners and ritually rejuvenated".[54] Just as the married daughter can gain succor and confidence from her mother upon returning home, renewing a link that may have been weakened by distance and time and by dwelling in another household, so too the secondary Mazu temples abroad can strengthen their religious legitimacy and authority through the pilgrimages to the Meizhou temple.

Second, "returning to mother's home" is also a metaphor for the Chinese diaspora, retrieving its common cultural roots and establishing community identity. The trip to visit the "Meizhou mother ancestor" (*Meizhou mazu*) is a journey back in space and time. Not only is it as if the married daughter who left the mother's home for a distant marriage comes back, but also the pilgrimage signifies a return to a primordial place, a place where an extended spiritual family originated. The pilgrimage made by the Chinese diaspora is a reunion with the mother culture and a restoration of their common memory as both Chinese and Mazu devotees. Sangren puts it well: Mazu has the maternal power to bring together all the children in the family, producing "an inclusive effect among Taiwanese

53 (Yang 2004, p. 224); also see (Guo 1993, p. 95). According to Nyitray Vivian-Lee, Taiwanese pilgrims did sense more greed than devotion. Meizhou locals, the temple associations, along with the PRC government did expect financial support from Taiwanese temples and pilgrims through donations of golden lotuses, golden plaques, and heavy red envelopes filled with cash. See (Nyitray 2000, p. 170).

54 (Sangren 1983, p. 9).

(uniting otherwise competitive Hakka, Chang-chou, and Ch'uan-chou factions)".[55] Beyond that, the Meizhou mother has the power to unify all devotees throughout the world and to establish a shared transnational identity among them. As I indicated in the introduction, I follow Linda Basch's and her colleagues' pioneering definition of transnationalism, which is "the processes by which immigrants forge and sustain multi-stranded social relations that link together their societies of origin and settlement. We call these processes transnationalism to emphasize that many immigrants today build social fields that cross geographic, cultural, and political borders".[56] In the last two decades, the literature on transnationalism has grown exponentially, but at the heart of this approach is a pointed critique of "methodological nationalism," the dominant perspective in the social sciences and humanities that takes the nation-state as a static, unified and given unit of analysis.[57] Without denying the continued centrality of the nation and national identity, the literature on transnationalism places the nation within multiple spatial scales, from the local to the global, that shape it and to which it creatively responds.

Adopting a transnational approach enriches our understanding of the cult of Mazu.[58] Through their pilgrimage to the Meizhou ancestral temple and their participation in powerful and highly emotive performances of a culture that is deemed foundational, the Chinese diaspora construct a common identity and a sense of belonging that transcend their geographic and political boundaries. As Yang points out, "it would seem that cross-strait Mazu pilgrimages are creating a regional ritual space and religious community of Chinese coastal peoples that do not conform to existing political borders".[59]

The mix of traditional and newly-created rituals attract over 200,000 Chinese people living abroad to Meizhou temple annually.[60] Since 1989, when the first wave of Taiwanese pilgrims undertook their "presenting incense" trip to Meizhou, the ancestral temple has organized a variety of events to promote cultural, tourist, and commercial exchanges between overseas Chinese and mainland China. For example, the ancestral temple organized academic and cultural exchange conferences between Taiwan and mainland China.[61] In terms of the tourist exchanges, the ancestral temple organized a photography exhibition with the theme of "Affection and Beauty of Meizhou Island," which showcased thousands of photographs from artists across the Taiwan strait. To celebrate the approval of Putian city as one of the tourist ports that is eligible to make direct trips across the Taiwan strait, the ancestral temple invited over 50 Mazu temples in Taiwan and 7000 Taiwan devotees to undertake a four-day pilgrimage. During the 10th Mazu Culture Tourist Festival, the event "Mazus in the world returning to mother's home" (*tianxia mazu hui niangjia*) attracted over 300 Mazu statues of affiliated temples. The commercial exchange is attested to by the "Mazu and Health" forum across the Taiwan strait which promotes the business cooperation between Putian and Taiwan companies. As a result, 18 business contracts worth a total of 7 billion dollars were signed.

By emphasizing its status as the birthplace of Mazu's worship and as a symbol of cultural integration among the Chinese people in the mainland China and abroad, the new approaches adopted by the Meizhou temple have successfully attracted affiliated temples

55 (Sangren 1983, pp. 15–16).
56 (Basch et al. 1994, p. 8).
57 See Wimmer and Schiller (2003) and Levitt and Glick-Schiller (2004).
58 For more on transnational approaches to Mazu worship, see Lee (2009).
59 (Yang 2004, p. 261). According to Yang, the majority of Taiwanese followers feel compelled to make pilgrimages to Meizhou temple since the Mazu cult originates from mainland China. This fact makes it more difficult for the Taiwan government to "appropriate the powerful Mazu cult as the ground for new Taiwanese nationalism." In this sense, Taiwanese pilgrims express their willingness to cross the political borders between Taiwan and the mainland. I agree with Yang that the unifying efforts of pilgrimage does not represent the political unification but religious unification of Mazu's devotees. The pilgrimages to Meizhou provide a public arena through which Taiwanese and mainland believers share their religious experiences and establish a common identity as Mazu's devotees).
60 This statistic comes from the temple magazine, *Mazu Homeland: Origin, Heritage, Circulation, Integration*, p. 38; see also Jiang and Zhu (2011, p. 120). See http://www.putian.gov.cn/zjpt/qhgk/xzqh/201210/t20121031_125384.htm. Accessed on 17 February 2021. This statistic is also confirmed by the Putian People's Government (2020).
61 The detailed information of various events organized by the Meizhou ancestral temple comes from the temple magazine, *Mazu Homeland: Origin, Heritage, Circulation, Integration*, pp. 38–40.

and overseas devotees, in particular Taiwanese devotees to undertake a pilgrimage to their ancestral and spiritual home.

The traditional ritual of "dividing incense" is crucial in building a transnational cultural community. The annually "returning to mother's home" pilgrimage trip serves to renew the sense of belonging that stretches the bounds of the nation to include all those places where Chinese immigrants live. Tourism has played a major role in this unbounding of Mazu, from an imperial metaphor to a trope of global identity and belonging.

5. The Actors Underlying the Development of Religious Tourism in Modern China

The transformation of Mazu worship into religious tourism in modern China is not simply a product of changing dynamics in "the religious" and "tourism" fields. At a deeper level, it is a sign of radical changes in the larger sociopolitical environment. In this section, I will discuss the political and economic agencies that led to this transformation. Post-1978, mainland China witnessed a wave of reinventions and innovations of old religious traditions that have adapted them to a modern context. This wave was the product of a new religious policy adopted by Chinese governments, as well as changes in official discourses regarding the legitimacy of religions, the impact of market economics, and the actions of local government to boost the local economy. Let me take up each of these dynamics.

The revival of popular religious traditions, including Mazu's cult across southeast China, was in large part made possible by a dramatic shift in the government's attitudes to Chinese religions. Before the enactment of the new constitution and Document 19, all religions had to contend with the dominant ideology, Marxism, which defines religion as "the opium of the people." With this understanding of religion, the Communist government asserted its supervision over religions by implementing religious policies that restricted the activities of religious institutions, priests, and believers. Popular religious practices were tagged as superstition, which was not protected by the law. However, the revision of the constitution adopted in 1982 and the implementation of "Document 19" that circulated in the same year ushered in a new period of religious tolerance. For instance, Document 19 states that "the basic policy the Party has adopted toward the religious question is that of respect for and protection of the freedom of religious belief".[62] This religious liberty clause provides official justification for a restoration of popular religions and the protection of the freedom of individual religious belief. Along with the policy of religious tolerance, "official religious associations were reinstated, officially designated places of religious worship were reopened, and religious communities were allowed and even encouraged to engage in international exchanges with their coreligionists".[63] This religious policy also distinguishes normal religious life from other illegal activities tagged as superstitions or evil cults. Normal religious activities permitted by the government and law include scripture chanting, ritual performance, self-cultivation, publishing religious books, and selling religious products. Religious activities, such as "divination, sorcery, exorcism, spirit procession, and fengshui," were categorized as superstitions and illegal.[64]

One policy in Document 19 emphasizes the tourist significance of some sacred sites with long history:

Churches and temples located at famous mountains and scenic resorts are not only religious sites, but also facilities with high historical and antique values. These churches and temples should be carefully maintained, the antiquity should be well preserved, the architecture should be properly renovated, the environment should be fully protected. In this way, these religious sites will become tourist sites with clean, peaceful, and beautiful environments.[65]

---

62 The English translation of Document 19 comes from MacInnis (1989, p. 10).

63 See (Goossaert and Palmer 2011, p. 324). They give a detailed analysis of Document 19 and its impact on religious practices in modern China.

64 See Document 19 on the Website of the "Chinese National Religions": (The Website of Chinese National Religions 1982) http://www.mzb.com.cn/html/folder/290171.htm (accessed on 10 July 2009); for information in English, see Goossaert and Palmer (2011, p. 325).

65 Ibid.

In addition to the changing religious policy, another political factor that has contributed to the revival of Chinese popular religious traditions is the emergence of other channels for institutional legitimation, such as the Chinese government's strong support for intangible cultural heritage. In 2004, PRC governments issued a decision to support the "Convention for the Safeguarding of the Intangible Cultural Heritage," which was first promulgated by The General Conference of the United Nations Educational, Scientific and Cultural Organization (UNESCO) in 2003.[66] "Intangible cultural heritage," as defined in the convention, comprises the following domains: "oral traditions and expressions, including language as a vehicle of the intangible cultural heritage; performing arts; social practices, rituals and festive events; knowledge and practices concerning nature and the universe; traditional craftsmanship".[67] On the basis of the UNESCO convention, the Chinese government issued "Suggestions for Reinforcing Our National Efforts to Safeguard the Intangible Cultural Heritage," which provides a detailed account of administrative procedures to implement the convention at a national level. According to the suggestions, the Ministry of Culture, together with provincial and county level cultural affairs bureaus, takes the responsibility to evaluate whether the items of traditional culture are qualified to be registered as intangible heritage. Once these items have been designated as such, they will benefit from special legal protection and financial support. This policy has opened a legal channel for popular religious practices to gain official recognition from central and local governments. As it happened during imperial China, recognition from the central government translates into legitimacy and authority, which, in turn, has served to increase the visibility of these practices and organizations and to attract more economic resources, including tourism. Thus, religious associations, in particular cults to popular deities and local ritual traditions, have enthusiastically devoted themselves to applying for the registration as intangible heritage. Now, major local cults to deities, such as Qingshui laozu, Mazu, and some cults to civilizing heroes, including Yu the great, Confucius, and the Yellow Emperor, are all inscribed in the list of cultural heritage protection. In 2009, UNESCO decided to inscribe Mazu worship on the list of cultural heritage protection for several reasons. First, Mazu worship is recognized by different social groups as a symbol of identity and is passed on as a continuous tradition from generation to generation. Second, the inclusion of Mazu worship into the list of "Intangible cultural heritage" promotes cultural diversity and human creativity. Third, actors at every level, from local temples and provincial and county level cultural affairs bureaus to the central government, are engaged in the promotion of the worship of Mazu, highlighting the significance of the cult. Fourth, the Mazu worship has been recognized by the Chinese Ministry of Culture as worthy of being included into the national list of "Intangible cultural heritage".[68]

In line with the policy of religious tolerance adopted by central government, local governments have also become important agents in the development of religious tourism. Specifically, local governments are actively involved in restoring and promoting religious institutions in their jurisdictions to bring income from tourism, as well as supporting the application of intangible cultural heritage to benefit from special funding by central government. In the case of Mazu worship, the government of Fujian province and the local government of Putian city where the ancestral temple is located have been at the forefront in the promotion of tourism to the temple. In fact, officials from the Fujian government and the Putian government actively participate in the Meizhou's Mazu Cultural Tourist Festival. As described in the local government document entitled "Construction Planning of the Ancestral Temple", the local government sponsored the restoration of Meizhou ancestral temple in 1980s in order to boost the development of local tourism.

Beyond political factors, the market economy has also played a significant role in the revitalization of Chinese popular religious traditions. As I hinted above in discussing

[66] "Convention for the Safeguarding of the Intangible Cultural Heritage": (The Website of Chinese Intangible Cultural Heritage 2003) http://www.ihchina.cn/zhengce_details/11667 (accessed on 11 July 2020).

[67] Ibid.

[68] See (Jiang and Zhu 2011, p. 137).

how official recognition has translated into increased public visibility, the expansion of the market economy in China has opened more opportunities for temple associations to increase their incomes. Specifically, the operation, and even survival, of temples in a market economy are increasingly dependent on donations and income generated by tourism and the provision of ritual services. In this dramatic changing economic context, temples have shifted from having religious associations to becoming socioeconomic entities. For example, Shaolin temple now operates as a commercial complex that provides all kinds of Kungfu performances as commodities to attract tourists. This has allowed it to transform itself from a Buddhist temple to one of the most popular tourist sites in China. The abbot Shi Yongxin has been referred as the "CEO monk".[69] Similarly, we saw how the association at Mazu's ancestral temple has become more differentiated and institutionalized to serve non-ritual functions. To attract more donations from devotees, temples now have to provide commodified ritual services and religious goods, as well as to organize activities such as religious tourism which were not traditionally connected with worship.[70]

Last but not least, the renewal of popular Chinese religion and, in particular, the national and transnational spread of the devotion to Mazu, has been aided by the increasing effort to reunify mainland China and Taiwan. The late 1980s and 1990s, large-scale pilgrimage activities across the strait was witnessed. The central government has encouraged temples and devotees from overseas to rebuild their spiritual connection with the ancestral temples in mainland China, retrieving their lost cultural roots and building a common religious community. The construction of a shared sense of spiritual and cultural belonging dovetails with the government's desire to promote China's political reunification and to project its influence globally.[71] In line with the mainland state agencies, the Meizhou ancestral temple and other original temples of popular deities have reshaped themselves as symbols of cultural exchange and common community.

6. Conclusions

The changing religious policy, and in particular, the emphasis of religious tourism, intertwined with the promotion of the registration of intangible heritage, the influence of market economy, as well as the role of Mazu in unifying overseas Chinese people, have contributed to the modern transformation of her worship. In this article, I have highlighted three dynamics that underlie these sociopolitical, economic, and religious changes: the invention of tradition, folklorization, and commodification. (Trans)national religious tourism has played a key role in producing these three dynamics. In order to maintain its hegemony as the preeminent temple for Mazu's devotion, the ancestral temple at Meizhou Island has reconstructed narratives and recovered rituals that make it the legitimate, originative, and most authentic expression of the cult. Moreover, in response to changes in policy by the central government, global understandings of heritage culture, and to the increasing penetration of a market economy, the temple association has transformed the Meizhou site architecturally and ritually. Sacred spaces and rituals that were originally connected to a small, local devotion have been given aesthetic–entertainment functions that fulfill the needs and desires of (religious) tourists, as well as a cultural valence, as an uncontested and authentic symbol of a core Chinese historic-cultural identity, which serves to generate a powerful sense of belonging not only for the Chinese in the mainland but, more importantly, for those in the diaspora. In that sense, Mazu is no longer "an imperial metaphor." She has become a trope for a sense of Chinese national identity that now circulates globally and sustains and is sustained by transnational religious tourism. Just as merchants, sailors, and government officials, both local and imperial, were key to spreading the cult of Mazu in pre-modern China, so too are transnational pilgrims,

69 For more on the commercialization of Shaolin temple see Zhao (2017).

70 Here commodified ritual services refer to religious rituals performed for commercial profit. Ritual services have been modified from their original form and function to appeal to and fulfill the interests and needs of tourists who pay established fees for these services. In turn, some of the profits collected through these services are used to promote more tourism and continue to increase profits.

71 See (Nyitray 2000, p. 170); (Goossaert and Palmer 2011, p. 261).

immigrants, and tourists today, as well as the PRC's administration and the global market of heritage culture, central in the transnational "unbounding", to refer back to Basch's approaches to Mazu (Basch et al. 1994). The transformations in Mazu's ancestral temple at Meizhou, and more generally in her cult, illustrate some of the most significant changes that the Chinese religious field is undergoing, as the country responds to global and transnational processes, including migration and tourism.

Funding: This research received funding from Fujian Provincial Humanities and Social Sciences Foundation (Fujian sheng sheke) No. FJ2020B136.

Institutional Review Board Statement: Ethical review and approval were waived for this study.

Informed Consent Statement: Informed consent was obtained from all subjects involved in the study.

Data Availability Statement: The data presented in this study are available on request from the corresponding author.

Conflicts of Interest: The author declares no conflict of interest.

References

Basch, Linda, Nina Glick Schiller, and Christina Szanton Blanc, eds. 1994. *Nations Unbound: Transnational Projects, Postcolonial Predicaments, and Deterritorialized Nation States*. London and New York: Routledge.

Bremmer, Thomas S. 2020. *Consider the Tourist. The Wiley Blackwell Companion to Religion and Materiality*. Edited by V. Narayanan. London: John Wiley & Sons, pp. 187–206.

Chau, Adam Yuet. 2006. *Miraculous Response: Doing Popular Religion in Contemporary China*. Stanford: Stanford University Press.

Eliade, Mircea. 1959. *The Sacred and the Profane: The Nature of Religion*. Boston: Houghton Mifflin Harcourt.

Goossaert, Vincent, and David A. Palmer. 2011. *The Religious Question in Modern China*. Chicago and London: The University of Chicago Press.

Guo, Jinrun, ed. 1993. *Dajia Mazu Jinxiang*. Tai-chung: Taichuang County Cultural Center.

Hobsbawm, Eric, and Terence Ranger. 1983. *The Invention of Tradition*. Cambridge and New York: Cambridge University Press.

Hung, Ying-Fa. 2009. *From Religious Pilgrimage to Tourism and Bodily Cultivation: Taiwan Dajia Mazu Pilgrimage*. Taiwan: The International Conference for Mazu Study: The Forum for Mazu Popular Belief and Culture [Mazu guoji xueshu yantao hui: Mazu minjian Xinyang yu wenwu lunwen ji], pp. 287–328.

Jiang, Jilin. 1987. *The Overview of Buddhist Mountains and Daoist Abbeys: The Bibliography of Heavenly Mother. [Fosha Daoguan Zonglan: Tianshang Shengmu Zhuanji]*. Taiwan: Yelin Publication [Yelin chubanshe].

Jiang, Weiyan, and Fuzhu Liu, eds. 2007. *The Compilation of Mazu's Historical Material and Texts: The Volumn of Pomes*. Beijing: Chinese Archives Press, [*Mazu Wenxian Shiliao Huibian: Shici Juan*. Beijing: Zhongguo dang'an chubanshe].

Jiang, Weiyan, and Lihang Zheng, eds. 2007. *The Compilation of Mazu's Historical Material and Texts (the First Volume): The Volume of Inscriptions*. Beijing: Chinese Archives Press, [*Mazu Wenxian Shiliao hulbian Diyi Ji: BEIJI Juan*. Beijing: Zhongguo dang'an chuban she].

Jiang, Weiyan, and Jintan Zhou, eds. 2009. *The Compilation of Mazu's Historical Material and Texts (the Second Volume: The Volume of Writings*. Beijing: Chinese Archives Press, [*Mazu Wenxian Shiliao Huibian Dier Ji: Zhulu Juan*. Beijing: Zhongguo dang'an chuban she].

Jiang, Weiyan, and Hepu Zhu, eds. 2011. *The Record of Mazu at Meizhou*. Fujian: Local Gazetteers Publications [*Meizhou Mazu Zhi*. Fujian: Fangzhi chuban she].

Johnson, David George. 2009. *Spectacle and Sacrifice: The Ritual Foundations of Village Life in North China*. Cambridge: Harvard University Press.

Juan, Shan. 2010. Growing Interest in Religious Tourism. China Daily. Available online: http://www.chinadaily.com.cn/china/2010-11/19/content_11574548.htm (accessed on 17 February 2021).

Lamley, Harry. 1977. The Formation of Cities: Initiative and Motivation in Building Three Walled Cities in Taiwan. In *The City in Late Imperial China*. Edited by Skinner. Stanford: Stanford University Press, pp. 195–96.

Lanquar, R. G. 2011. Pilgrims between East and West. in Religious Tourism. In *Asia and the Pacific*. Madrid: World Tourism Organization (UNWTO), pp. 1–9.

Lagerwey, John. 2010. *China, a Religious State*. Hong Kong: Hong Kong University Press.

Lee, Jonathan H. X. 2009. Creating a Transnational Religious Community. In *Religion at the Corner of Bliss and Nirvana*. Edited by Lois Lorentzen. Durham and London: Duke University Press.

Levitt, Peggy, and Nina Glick-Schiller. 2004. Conceptualizing Simultaneity: A Transnational Social Field Perspective on Society. *The International Migration Review* 38: 1002–39. [CrossRef]

Li, Xianzhang. 1995. *The Study of Mazu Belief*. Macau: Macau Maritime Affairs Library Press [*Mazu Xinyang Yanjiu*. Macau: Macau haishi bowu guan chuban].

Liao, Chien Chih. 2019. *The Study on the Marketing of Mazu Culture Derivative Products*. The Collected Papers of the Fifth International Forum of Mazu Culture. Putian: Putian College, [*Mazu Wenhua Yansheng Wenchuang Shangpin Yingxiao Zhi Yanjiu. Diwu Jie Guoji Mazu Wenhua Xueshu yantao Hui Lunwen Huibian*. Putian: Putian xueyuan].

Liu, Dake. 2011. *The Tradition and Transformation: Fujian People's Popular Beliefs*. Beijing: The Social Science and Material Press [*Chuantong Yu Bianqian: Fujian Minzhong de Xinyang Shijie*. Beijing: Shehui kexue wenxian chuban she].

MacInnis, Donald E. 1989. *Religion in China Today: Policy and Practice*. Maryknoll: Orbis.

Norman, Alex. 2013. *Spiritual Tourism: Travel and Religious Practice in Western Society*. London: Bloomsbury Academic.

Nyitray, Vivian-Lee. 2000. Becoming the Empress of Heaven: The Life and Bureaucratic Career of Mazu. In *Goddesses Who Rule*. Edited by Beverly Moon and Elizabeth Benard. New York and Oxford: Oxford University Press, pp. 165–80.

Poceski, Mario. 2009. *Introducing Chinese Religions*. London: Routledge.

Putian People's Government. 2020. Meizhou Island. Putian Shi Renmin Zhengfu Meizhou Dao. Available online: http://www.putian.gov.cn/zjpt/qhgk/xzqh/201210/t20121031_125384.htm (accessed on 17 February 2021).

Raj, Razaq, and Kevin Griffin, eds. 2015. *Religious Tourism and Pilgrimage Management: An International Perspective*, 2nd ed. Wallingford: CAB International.

Rocha, Cristina. 2006. Spiritual Tourism: Brazilian Faith Healing Goes Global. In *W. H. Swatos, On the Road to Being There*. Leiden: Studies in Pilgrimage and Tourism in Late Modernity, pp. 105–23.

Rubinstein, Murray A. 1994. *The Revival of the Mazu Cult and of Taiwanese Pilgrimage to Fujian*. Cambridge: Harvard University, Fairbank Center for East Asian Research.

Sangren, P. Steven. 1983. Female Gender in Chinese Religious Symbols: Kuan Yin, Mazu, and the Eternal Mother. *Signs* 9: 4–25. [CrossRef]

Sangren, P. Steven. 1993. Power and Transcendence in the Ma Tsu Pilgrimages of Taiwan. *American Ethnologist* 20: 564–82. [CrossRef]

Schipper, Kristofer. 1990. The Cult of Pao-Sheng Ta-Ti and its Spreading to Taiwan: A Case Study of Fen-hsiang. In *Development and Decline of Fukien Province in the 17th and 18th Centuries*. Edited by Eduard B. Vermeer. Leiden: Brill.

Stausberg, Michael. 2011. *Religion and Tourism: Crossroads, Destinations and Encounters*. London: Routledge.

Tan, Yanhan. 2015. 2009–2012 The Review of Chinese Religion and Tourism from 2009 to 2012 zhongguo zongjiao lüyou yanjiu zongshu. The Study of Tourist Relationship. *Lüyou Guanli Yanjiu* 1: 26–27.

Ter Haar, Barend J. 1990. The Genesis and Spread of Temple Cults in Fukien. In *Development and Decline of Fukien Province in the 17th and 18th Centuries*. Edited by Eduard B. Vermeer. Leiden: Brill.

The Website of Chinese Intangible Cultural Heritage. 2003. Convention for The Safeguarding of The Intangible Cultural Heritage. Available online: http://www.ihchina.cn/zhengce_details/11667 (accessed on 21 August 2020).

The Website of Chinese National Religions. 1982. Document 19 of Chinese National Religions. Available online: http://www.mzb.com.cn/html/folder/290171.htm (accessed on 10 July 2020).

Timothy, Dallen, and Daniel Olsen, eds. 2006. *Tourism, Religion & Spiritual Journeys*. New York: Routledge.

Tischer, Jacob Friedemann. 2018. Mazu Nation: Pilgrimages, Political Practice, and the Ritual Construction of National Space in Taiwan. *Global Politics Review* 4: 6–28.

Trouillot, Michel-Rolph. 1995. *Silencing the Past: Power and the Production of History*. Boston: Beacon.

Vasquez, Manuel A., and Knott Kim. 2014. Three Dimensions of Religious Place Making in Diaspora. *Global Networks* 14: 326–47. [CrossRef]

Wang, Qingsheng. 2018. Empirical Study and Inspiration of Taiwan Mazu Folk Sports Industrialization: Taking the Dajia Mazu International Tourism Cultural Festival as an example. *Mazu Wenhua Yanjiu* 4: 76–88.

Watson, James L. 1985. Standardizing the Gods: The Promotion of T'ien Hou (Empress of Heaven) along the South China Coast, 960–1960. In *Popular Culture in Late Imperial China*. Edited by David Johnson, Andrew J. Nathan and Evelyn S. Rawski. Berkeley and Los Angeles: University of California Press, pp. 292–324.

Wimmer, Andreas, and Nina Glick Schiller. 2003. Methodological Nationalism, the Social Sciences, and the Study of Migration: An Essay in Historical Epistemology. *The International Migration Review* 37: 576–610. [CrossRef]

Yang, Mayfair Mei-hui. 2000. Putting Global Capitalism in Its Place Economic Hybridity, Bataille, and Ritual Expenditure. *Current Anthropology* 61: 477–509. [CrossRef]

Yang, Mayfair Mei-hui. 2004. Goddess across the Taiwan Strait: Matrifocal Ritual Space, Nation-State, and Satellite Television Footprints. *Popular Culture* 16: 209–38. [CrossRef]

Zhao, Dong. 2017. Transformation of the sacred: The Commercialization, Politicization and Globalization of Buddhist Heritage Tourism in Contemporary China. *International Journal of Business Tourism and Applied Sciences* 5: 44–52.

Zhang, Yanchao. 2019. The State Canonization of Mazu: Bringing the Notion of Imperial Metaphor into Conversation with the Personal Model. *Religions* 10: 151. [CrossRef]

Zheng, Lihang, ed. 2011. *The Compilation of Mazu's Historical Material and Texts (the Third Volume): The Volume of Local Gazetteers*. Fujian: Hifeng Press [*Mazu Wenxian Shiliao Huibian Disan ji: Fangzhi Juan*. Fujian: Haifeng chuban she].

Article

Religious Tourism's Impact on City Space: Service Zones around Sanctuaries

Izabela Sołjan and Justyna Liro *

Institute of Geography and Spatial Management, Jagiellonian University, 30-387 Kraków, Poland; izabela.soljan@uj.edu.pl
* Correspondence: justyna.liro@uj.edu.pl

Abstract: Pilgrimage centers are important elements of the spatial structure of cities and simultaneously factors influencing their transformations. The pilgrimage function of sanctuaries can lead to development of service zones around them focused mainly on serving visitors, i.e., pilgrims and tourists. They often perform functions complementary to sanctuaries. Here we present the results of studies of sanctuary service zones conducted at twenty six popular Catholic sanctuaries in Europe. In this paper, we discussed the influence of the sanctuary on city space on the macro, meso and micro scales. We proposed a definition of a sanctuary service zone, and developed a model approach to the different types: initial (slightly developed) zones, dispersed zones (integrated into the urban space, with their pilgrimage function coexisting with other urban functions), and compact zones—linear, or pilgrimage districts (with dominating pilgrimage function). The development of sanctuary service zones depends mainly on the rank of the pilgrimage center, as well as on the period in which it was founded, pilgrimage traditions, and the location of the pilgrimage center in the city. This paper is a continuation and extension of research into the impact of pilgrimage centers on city space transformations in the context of socio-cultural changes in the 20th and 21st centuries.

Keywords: pilgrimage center; sanctuary; sanctuary service zones; pilgrimage; religious tourism; city space; spatial development; tourist function; sacred sites; geography of religion

Citation: Sołjan, Izabela, and Justyna Liro. 2021. Religious Tourism's Impact on City Space: Service Zones around Sanctuaries. *Religions* 12: 165. https://doi.org/10.3390/rel12030165

Academic Editor: Kiran Shinde

Received: 1 February 2021
Accepted: 2 March 2021
Published: 4 March 2021

1. Introduction

Religious tourism is one of the fastest-growing, changing sectors in the world tourism market (Reader 2014; Eade 2017; Collins-Kreiner 2019) and religious motives are still one of the common reasons for travelling around the world. As a result of socio-cultural changes in the 20th and 21st centuries—for example, mass tourism development (including religious tourism); the universality, accessibility and growing popularity of travel; the secularization, commercialization (Kaufman 2005) and globalization of societies and life (Gökariksel 2009)—changes in pilgrims' needs, motivations and behavior have occurred (Sołjan and Liro 2020; Liro 2021). Social progress has also affected the ways in which people go on pilgrimages. They demand higher-quality accommodation and other features of modern life also in pilgrimage centers (Liro et al. 2018; Sołjan and Liro 2020). Their expectations and new standards of facilities result in increasingly extensive, diverse and spatial development sanctuaries (Sołjan 2012; Sołjan and Liro 2020) and significantly influence the spatial structures of the cities where they are located. All these changes might be a response to the needs of the 'new' pilgrimage center visitors mentioned by Eade and Sallnow (1991); Badone and Roseman (2004); Collins-Kreiner (2010a, 2010b), Shinde and Rizello (2014) and (Liro 2021; Liro et al. 2017), whose motivations and expectations become diverse and have much in common with tourism in the traditional sense.

Despite this global importance and growing popularity of religious tourism, there is still a lack of knowledge about religious tourism's impact on city space. Due to the spatial-temporal influence on geographical space, religious tourism is an important research issue, especially in tourism geography, the geography of religion and urban studies

(Deffontaines 1948; Fickeler 1947; Holloway and Vallins 2002; Shinde and Olsen 2020; Sopher 1981). Travel to sacred places was considered as a factor sacralizing space (Fickeler 1947), influencing the development of towns and settlements (Deffontaines 1948), as well as significant factors changes in the socio-geographical environment, i.e., changes in settlement networks, or cultural landscape transformations (Sopher 1981). Recently, numerous studies have pointed out the strong need to supplement research on the interaction between pilgrimages and geographical space (Collins-Kreiner 2010a, 2010b; Holloway and Vallins 2002; Rinschede and Sievers 1987) because, apart from mentioned below papers, there is no theoretical basis geographical research on the impact of the development of religious tourism on the transformation of spatial structures in cities.

Pilgrimage centers have been the subject of numerous studies, but there are very few theoretical papers about their spatial development and their influence on city space (Liro 2018; Liro and Sołjan 2016; Nolan and Nolan 1989, 1992; Rinschede 1987, 1988, 1995; Ambrosio 2007; Ambrosio and Pereira 2007; Sołjan 2011, 2012; Sołjan and Liro 2015, 2020). The network of 6150 pilgrimage centers functioning in Western Europe was described by Nolan and Nolan (1989). The sanctuary's influence on a town's development (in the macro-scale approach) and its infrastructure was described through the examples of Lourdes, Fatima, and Loreto (Rinschede 1987, 1988, 1995). Rinschede (1987, 1988, 1995) distinguished three sanctuary development factors, on the basis of which research should be conducted: (1) spatial structure, (2) its social impact, and (3) pilgrimage movement range. Ambrosio (2007) and Ambrosio and Pereira (2007) analyzed the development of four pilgrimage centers in Europe (Banneux, Fatima, Knock and Lourdes) in comparison with the development of urban units. They highlighted the importance of place marketing, promotion and spatial development planning for the successful development of the pilgrimage center and the city itself, as well as the type of land ownership for the development of service infrastructure in the immediate vicinity of the pilgrimage center. The spatial changes were also shown in contributing papers based on examples of selected pilgrimage centers in Poland (Liro 2018; Liro and Sołjan 2016; Sołjan and Liro 2015).

The mentioned socio-cultural changes in the 20th and 21st centuries and the blurring of the boundaries between tourism and pilgrimages in many aspects have influenced pilgrimage centers (Sołjan 2012; Sołjan and Liro 2020). Today, pilgrimage centers are developed with multiple buildings (called sacred complexes or religious-recreational parks), offering additional cultural and tourist facilities for visitors. Contemporary pilgrimage centers are diverse in size, spatial organization, the number and type of buildings, as well as in the their development (Sołjan 2012; Sołjan and Liro 2020). Based on studies on pilgrimage centers, and taking into account the theoretical backgrounds to the space in sanctuaries (Rinschede 1987, 1988, 1995), a model of the contemporary pilgrimage center was developed (Sołjan and Liro 2020). The model approach on the basis of empirical research assumed that the area around the sanctuary can be developed into a service zone (Figure 1). The service zone is an area of high activity for visitors, transformed by their functions (Sołjan 2012; Sołjan and Liro 2020). The service zone is aimed mainly at satisfying the material and tourist needs of pilgrims, influenced by the size of the pilgrimage movement.

Sanctuaries are visible in the structure of cities and are one of its many spatial elements (Sołjan 2012). Religious, architectural, and other values, for example cultural ones, make sanctuaries important objects in the city space, which arouse the interest of pilgrims and tourists. They can also be a factor actively affecting urban space, influencing its changes and transformations. Owing to the existing gap in knowledge about the impact and significance of the sanctuary for the development of urban space, this article is a continuation and extension of research into spatial development and the impact of sanctuaries on city space transformations in the context of socio-cultural changes in the 20th and 21st centuries. Particular attention was paid to *sanctuary service zones*, the definition of which was only introduced in the previous paper (Sołjan and Liro 2020).

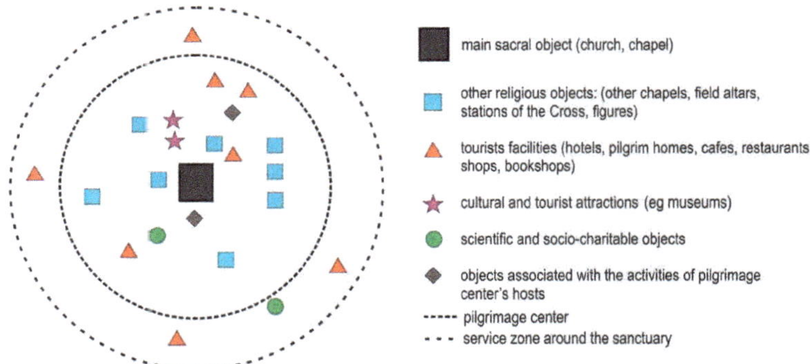

Figure 1. A model of the contemporary pilgrimage center (Figure 3; Sołjan and Liro 2020).

The aim of this paper is to present religious tourism's impact on city space based on the service zones around pilgrimage centers. Here, we present the results of studies conducted at twenty six popular Roman Catholic sanctuaries in Europe (Altötting, Assisi, Bardo, Czestochowa, Einsideln, Fatima, Kalwaria Wejherowska, Kalwaria Zebrzydowska, Kevelaer, Knock, Cracow (2), Lewocza, Loreto, Lourdes, Mariazell, Padua, Paris, Pompeii, Rome, San Giovanni Rotondo, Santiago de Compostela, Syracuse, Tuchow, and Zakopane). Particular attention was paid to the influence of sanctuaries on the urban space in the macro, meso and micro scales. We proposed a definition of a *sanctuary service zone*, and this paper is summarized using the model approach to sanctuary service zones types proposed here for the first time.

2. Materials and Methods

The materials on the impact of pilgrimage centers and the transformation of city space as well as data on the areas of visitor activity come from the field inventory carried out in the period 2008–2018, archival resources, and available studies. A valuable source of information was non-standardized interviews with the hosts of the centers. Some data were obtained from commune and municipality authorities. The research took place in the following way: in order to understand the structure of sanctuary service zones, cartographic and archival materials were analyzed. Then, a field inventory was conducted to collect current data on the areas of visitor activity and inventory of tourist infrastructure. Next, a comparative analysis of the results from the cartographic material analysis and the data obtained during the field studies was performed in order to determine religious tourism's influence on city space and develop a conceptual model of sanctuary service zones.

Field studies, inventories of objects related to the functioning of pilgrimage centers, and their location in relation to the core (main sacred object) of the sanctuary with the marking of their distance from the sanctuary service zone, and the spatial and historical analysis of the obtained data constitute the methodological basis of the presented article. The influence of pilgrimage centers on the organization of urban space has not been the subject of comprehensive geographic studies so far. In our research, we attempted to fill this gap by tracing the impact of pilgrimage centers on the organization of urban space, taking into account three spatial scales, from the immediate surroundings of the sanctuary to its impact on a scale of the entire town. The adopted criteria (described in this article) for each of the separated scales made it possible, on the basis of the conducted field inventory as well as spatial and historical analysis, to determine the impact of the sanctuary on micro, meso, and macro scales. The field studies were also aimed at separating and characterizing sanctuary service zones and showing their diversified impact on the city, from the micro scale (small zones) to the macro scale in the case of the largest pilgrimage districts. The areas closest to the sanctuary, associated with the highest number of visitors, were assumed to be sanctuary service zones. Earlier surveys on the area of the greatest activity of pilgrims, conducted by I. Sołjan in 2008, inter alia in Lourdes and Częstochowa, made it possible to determine the maximum range of these zones within a radius of 1000 m from the boundaries of the sanctuary, and so this range was adopted in this article. The main determinant of the zone was the presence of religious facilities and tourist infrastructure created to meet the needs of people visiting the pilgrimage center.

3. Results

3.1. The Sanctuary in the City Space

Sanctuaries, through their architectural form and functions, mark their presence in the urban space. They constitute a significant, symbolic space, often with great religious and cultural potential. Their impact on the city may be of a different nature and may have various effects (Figure 2). The conducted research on selected European Catholic sanctuaries has made it possible to distinguish three important spatial scales of the described phenomena (Sołjan 2012), (Figure 2):

MACRO

- city-forming factor,
- spatial development of suburban areas or new districts including large pilgrimage districts,
- changes in the way of development and functions of the existing city areas,
- strong development of tourist infrastructure,
- expansion of the communication network,
- reactivation of former cities.

MESO

- development of facilities: places related to people and events important to the sanctuary, museums and exhibitions, and other cultural institutions,
- establishment and greater concentration of religious institutions: religious associations and organizations with pilgrimage houses, social and educational institutions (also for residents),
- development of the tourist facilities,
- including other facilities in the area of pilgrims' activity,
- development of a general urban infrastructure,
- development of a specialist infrastructure, e.g. for the sick and the disabled.

MICRO

- development relatively small zone with tourist infrastructure (accommodation and catering facilities, and commercial establishments mainly shops with devotional items and souvenirs),
- presence of religious objects (other than in sanctuary), most often chapels,
- presence of religious orders and other religious associations' seats, often providing services also for pilgrims,
- development of infrastructure intended for sick and disabled pilgrims,
- religious nomenclature of the streets, squares, and facilities occurring in it.

Figure 2. The spatial scales of pilgrimage centers and religious tourism's impact on the city space.

I. Changes on a **macro** scale: these are the greatest changes taking place in the city under the influence of the pilgrimage center, leading to fundamental changes in the urban structure and its functions.

II. Changes on a **meso** scale: the impact of the pilgrimage center on a meso scale is manifested in the creation of various objects, institutions, and places in the city space related thematically or functionally to it, which are located outside the sanctuary and its immediate surroundings.

III. Changes on a **micro** scale: the impact on a micro scale leads to changes in the urban space only in the immediate vicinity of the sanctuary. The effect of the discussed influence of the pilgrimage center is a relatively small zone with service infrastructure used by pilgrims and tourists developing around it.

3.1.1. Macro-Scale Impact on City Space

The greatest influence of a pilgrimage center on the organization of the space and structure of a city, defined as a macro-scale impact, takes place when there are serious spatial and functional changes in the urban structure, covering even entire settlement units in the case of smaller city centers, or significant parts of them (in the case of larger cities). Among these types of changes, the most important are changes:

- Resulting in the emergence of a city center;
- Leading to the spatial development of an existing city and the emergence of new districts, including large sanctuary service zones (pilgrimage district);
- In the way of development and functional changes in the already existing city areas;
- Causing the reactivation of former cities.

All these processes are generally long term, observed from a multiyear historical perspective, fundamentally changing the urban landscape. They occur mainly in the case of highly renowned (international and national) pilgrimage centers with several million people each year. They also depend on the stage of development of a given city, as well as on many social conditions, including urban policy, and sometimes even the policy of state authorities (the example of Fatima).

The city-forming function of pilgrimage centers was observed as early as in ancient times, in many now forgotten religions, such as Mesopotamia, ancient Egypt and Greece. In Western Christianity, the Middle Ages were a period favorable for the development of cities based on sanctuaries. The best example here is Santiago de Compostela, today a city with 100,000 inhabitants in Spanish Galicia, the origin of which was the tomb of St. James the Apostle discovered there at the beginning of the 9th century. Miraculous events such as this attracted the local population and aroused the interest of the church hierarchy, which resulted not only in establishing a holy place, but also in the development of settlements around it. Despite the change in social and political paradigms, the end of the Middle Ages did not stop these processes and new sanctuaries and new cities continued to emerge. Such aspirations were fostered, for example, by counter-reformation trends aimed at renewing and reviving Catholicism. In 1602, the first Calvary was founded in Poland, and the settlement which developed around the Calvary and the monastery was granted a town charter as Nowy Zebrzydow as early as in 1640. The current name of Kalwaria Zebrzydowska was not adopted until the 19th century. Additionally, in Poland, the establishment of Kalwaria Wejherowska in Pomerania was associated with the simultaneous location of the city of Wejherowo, which was granted a city charter in 1648. In modern times, despite the basic formation of the settlement network in many European countries, it also happened that a religious center was the factor behind the foundation of a village, and then a city. Fatima is worth mentioning here, established in 'cruda radice' around the site of Our Lady's revelations in 1917. This small city has served 7–9 million pilgrims and tourists annually in recent years.

Under the influence of the pilgrimage center, a strong transformation of the spatial structure of already existing cities may also take place. The emergence of a sanctuary at a certain stage of a town's development may cause far-reaching changes in the organization of the center's space, as long as it is a dynamically developing sanctuary and its impact extends well beyond the borders of the region, or even the country. It is worth mentioning here the example of Lourdes, which was a little known town before the Marian revelations in 1858, serving as a local administrative and service center. As the result of the actions of the sanctuary, church policy, and widely publicized information about the revelations, both by the church and the media, already 50 years after the first revelations it was the most important Marian pilgrimage center in the world, visited by hundreds of thousands of pilgrims. The pilgrimage function of the city, which was also backed by the municipal authorities, caused a strong transformation of the urban structure, including the demolition of a number of houses and the construction of a new communication artery leading to the sanctuary. On previously undeveloped areas, the so-called lower town, a large pilgrimage district, the Grotto zone, was built with a strongly developed tourist infrastructure intended for visitors to the place of revelations. We observe a similar scenario of the development of San Giovanni Rotondo, a city famous for the sanctuary of St. Padre Pio, one of the most popular saints today. The expansion of the tourist infrastructure here runs in two ways, i.e., both in the historic part of the city and in the area of the pilgrimage center, where territorial resources make it possible to construct numerous buildings, mainly service facilities. Since the day of Padre Pio's canonization in San Giovanni Rotondo, the construction of 120 hotels has started, which shows the scale of these transformation processes (Giovine 2015).

A sanctuary may not only cause the creation of new areas in a city, but also change the development and functions of the existing districts. Highly renowned pilgrimage centers may displace other functions of the district, in which the pilgrimage function becomes the most important. This scenario generally concerns large, multifunctional cities, such as Cracow. In one of the districts of Cracow, in Podgorze, or more precisely in Łagiewniki and Borek Fałęcki, we currently have two large sanctuaries of international range, i.e., the Divine Mercy Sanctuary and the John Paul II Centre, still under development. The emergence of this sacred complex, and especially the strong development of the pilgrimage movement since the 1990s, has contributed to a change in the nature of the district, currently performing the pilgrimage function as the leading one. Following this, appropriate communication solutions are introduced, fundamentally changing the solutions in this area in the district, and thus in the city. Only commemorative plaques remind us of the old quarries and factories once present in these areas.

The city reactivation as a result of pilgrimage center is extremely rare. In this context, New Pompeii with the sanctuary of the Queen of the Holy Rosary can be mentioned. The ancient city was completely destroyed after the eruption of Mount Vesuvius in 79 AD. At the end of the 19th century, as a result of the missionary activity of Bl. Bartolo Longo, New Pompeii began to emerge from the surrounding settlements. Striving for the religious revival of the local population, he promoted the cult of Our Lady of the Rosary. Residential infrastructure and other public utility buildings developed around the church built from social donations. In 1927, New Pompeii received the status of an independent commune.

3.1.2. Meso-Scale Impact on City Space

The term meso-scale impact is defined as the presence in a city of places and institutions that are thematically or functionally related to the pilgrimage center, located in other parts of the city, outside the zone of direct impact of the sanctuary (Figure 2). These can be sacred buildings, places related to people and events important for the sanctuary, museums and religious exhibitions, as well as tourist facilities. Institutions functionally related to the pilgrimage center are primarily religious associations and organizations. This type of impact is not common and is more often observed in small urban centers which are large pilgrimage centers at the same time. In Fatima, approx. 2 km from the sanctuary, in Aljustrel—the home town of the visionaries Lucia, Francisco, and Jacinta)—a sacred and service zone complementary to the pilgrimage center was created. In addition, the monumental, so-called Great Stations of the Cross were established. In Lourdes, apart from the pilgrimage district clearly formed around the sanctuary, two large complexes serving pilgrims were established, i.e., the town of Cité Saint Pierre and the Youth Village. Cité Saint Pierre started to function in 1995 on an area of 32 ha. Six infrastructure buildings were built there, which can accommodate 500 people at a time. The Youth Village was built in 1988, in the middle of the forest, on the site of a former French Scout camp. Every year, several thousand young people from many countries of the world stay here.

The presence in the urban space of objects (or their complexes) thematically related to the sanctuary is important from the point of view of the development of the city and the pilgrimage center. It can lead to the activation of pilgrims, and thus enlarge their activity in the city. This, in turn, may increase the duration of pilgrims' stay and, taking into account the multiplier effects, it may increase the city income and the state of the economy. That is why the influence of a pilgrimage center on a meso scale is so important for smaller cities, which primarily perform the pilgrimage function. The impact on a meso scale depends not only on a sanctuary's hosts, but also on the activity of other entities, especially church and city authorities. In recent decades, owing to the development of religious tourism, many cities, including large ones, have marked out tourist routes based on religious potential and linked them to sanctuaries. In San Giovanni Rotondo, the city authorities took the initiative to attract pilgrims from strictly the area of the sanctuary to the historic city center. For this purpose, since the 1990s, the churches located in it began to be restored and promoted, emphasizing not only their religious and historical values, but above all their connections with St Padre Pio. In Lourdes, as part of the celebration of the 150th anniversary of the revelations, a special thematic path 'In the Footsteps of St Bernadette', and in Cracow, in order to arouse interest of those coming to the Divine Mercy Sanctuary and the John Paul II Centre in other tourist areas of the city, the route of St Faustina and the route of St John Paul II were created.

Pilgrimage centers may also affect the location of tourist facilities in parts of the city. Among the visitors to pilgrimage centers, there are also those who are looking for silence and want to spend time in concentration, away from the bustling zone filled with numerous hotels, restaurants, and shops.

Another effect of the impact of a pilgrimage center on a city on a meso scale may be a greater concentration of religious institutions, mainly religious congregations and associations. Some of the monasteries run retreat houses and provide accommodation services for pilgrims (e.g., in Fatima, Czestochowa, Cracow, Tuchow, and Zakopane). The presence of monastic facilities also favors the creation of educational institutions run by the church in the city (e.g., in Pompeii). Thus, the impact of a pilgrimage center on a city can be multifaceted, directed not only at the pilgrims arriving at the city, but also directly on its residents (thus, it is also an endogenous impact).

The presence of a pilgrimage center in a city may also lead to the development of infrastructure related to a specific group of pilgrims. In recent years, most attention—not only in pilgrimage centers—has been given to the sick and the disabled, and it results in adapting, e.g., railway stations, airports, or parking lots.

3.1.3. Micro-Scale Impact on City Space

Changes on a micro scale are the most common changes resulting from the existence and development of a pilgrimage center. Their effect is the development of a service zone around the sanctuary, adapted to serve pilgrims. However, this zone is relatively small, with a poorly developed tourist infrastructure, limited to a few or a dozen or so facilities (Figure 2). It is usually located in the streets or street adjacent to the pilgrimage center. We observe such a model of impact of a pilgrimage on a city in the following cases:

- In small sanctuaries, most often of regional range;
- Accumulation of tourist infrastructure within the pilgrimage center;
- As a certain transitional stage in the development of the pilgrimage center and the sanctuary service zone;
- Lack of adequate territorial facilities in the vicinity of pilgrimage center;
- Low activity of the entities administering the pilgrimage center and city, the residents.

Since the impact on a micro scale is limited mainly to the formations of sanctuary service zones, this issue will be discussed in detail in the manuscript's next sub-section.

3.2. Sanctuary Service Zones

The most visible manifestation of the impact of a pilgrimage center on the urban landscape is the emergence of a zone around it, serving pilgrims and tourists. It is primarily of a service and commercial nature, but can also be filled with objects of religious character. However, also in the case of the religious function of objects functioning in this zone, such as seats of religious congregations, they often provide tourist services for pilgrims, e.g., by providing accommodation. Historically, the zones in question have emerged for two main reasons:

1. Necessity to strictly separate the sacred and profane zones, justified by religious requirements;
2. Visitor needs other than spiritual or religious.

The separation of the sacred zone from the profane zone is related to the very idea of a sacred place, i.e., one dedicated to God (a deity), where appropriate rituals and rules of behavior apply. The separation of the sacred space took place both realistically by establishing the boundaries of the sanctuary, and symbolically by the performance of cleansing rites by the people arriving. Referring to the biblical description of Jesus's anger seeing merchants in the Jerusalem temple, for centuries attempts have been made to avoid any gainful activity in their area as unworthy of such places and offensive to God. The sphere of such activities was situated outside the walls of the pilgrimage center.

On the other hand, for the pilgrimage center visitors, especially arriving from afar, and spending a longer time in the city, a natural need was to provide a place to sleep, eat, and buy souvenirs. It favored the development of tourist and commercial facilities. However, it should be noted that even in the Middle Ages, stalls with devotional items, souvenirs, or jewelry were most often put up near the temple. Pilgrims spent nights away from the pilgrimage center, often on the outskirts. This was the case, for example, in Santiago de Compostela, where in the 10th century, owing to the growing importance of the sanctuary, urbanization processes in suburban settlements began (e.g., Lucio, Pinario). At the beginning of the 12th century, along with the construction of the new cathedral, the adjacent areas began to be developed, numerous shops and currency exchange points were built and, in the previously undeveloped areas, residential houses and facilities for pilgrims were constructed. This pattern was repeated over centuries in large pilgrimage centers, with urban development providing a commercial infrastructure expanding around the sanctuary.

However, the model that has been in force for centuries has undergone fundamental changes in modern times, especially since the 1980s. Undoubtedly, these changes were based on social transformations and, with them, patterns of religious devotion and the perception of a holy place (Sołjan and Liro 2020). Currently, in numerous pilgrimage centers, the presence in the strictly sacred zone of typical tourist infrastructure, with pilgrimage houses, gastronomic establishments, and commercial points, is increasingly often marked. This leads to inhibiting the development of sanctuary service zones, because their functions are taken over by pilgrimage centers. The problem is significant because in the described situation, it is not only about changing the entity generating economic profits, since it is also in sanctuary service zones that church authorities often owned many facilities, but about a paradigm shift, a qualitative change of the essence of the holy place and, in this context, the pilgrimage center. What has been forbidden and stigmatized so far for religious reasons and motives is slowly becoming a new norm. Commercialization enters pilgrimage centers increasingly from sanctuary service zones. Owing to the processes presented above, the formation of sanctuary service zones is not observed in the vicinity of many modern and dynamically developing pilgrimage centers, or their development is minimal.

Since so far in the state of knowledge, sanctuary service zones have not been defined or terminologically determined, we propose to assume that a sanctuary service zone is one that develops directly around the pilgrimage center, functionally complementary to it, developing within a radius of approx. 1000 m and performing primarily service functions for pilgrims and tourists.

The distinctive features of a sanctuary service zone include the presence of tourist infrastructure, which makes it possible to satisfy the needs of pilgrims. These are accommodation facilities, catering establishments, and commercial points (mainly shops with devotional items and souvenirs). Owing to the religious dimension of the pilgrimage center, sanctuary service zones are often also characterized by other features, which, however, are not necessary for their existence (Figure 2). These can include

- The presence of religious orders and seats of other religious movements and associations, often providing services also for pilgrims;
- The presence of museums and other cultural institutions, the subject matter and activities of which are related to the sanctuary;
- The presence of infrastructure intended for sick and disabled pilgrims;

- The religious nomenclature of streets, squares, and objects occurring in it.

An important prerequisite for the efficient functioning of sanctuary service zones is also the accessibility of pilgrimage centers in terms of communication accessibility and communication solutions inside them.

The designation of sanctuary service zones was based on two criteria:

a. The morphological structure of the area, including all service facilities and other objects and institutions serving pilgrims;

b. The range of the area of pilgrim activity, the area outside the pilgrimage center. This area was designated by drawing the isochrones of pedestrian access to the sanctuary along the main communication routes leading to it. The 15 min isochrone of walking to the pilgrimage center was adopted as the maximum (Sołjan 2012). Based on the conducted surveys and polls in the analyzed pilgrimage centers (Sołjan 2012), the maximum range of sanctuary service zones was assumed at 1000 m, and the service infrastructure for pilgrims in this area was analyzed.

The areas around pilgrimage centers can be developed in various ways and their nature can also be different. Sometimes, it is only a single object, e.g., a pilgrim's house, a shop with devotional items, or a restaurant. A noticeable regularity in sanctuary service zones is the increasing concentration of commercial services, especially shops with religious items, as they approach the pilgrimage center (Rinschede 1988, 1995). The size of the zones varies. They usually stretch along the main communication routes at a distance of 300–500 m (at most 1000 m) from the borders of the pilgrimage center (Sołjan 2012).

Before the characteristics of sanctuary service zones will be discussed, it is worth emphasizing that in the case of many pilgrimage centers, the above-mentioned zones are not formed at all and no changes are observed. Such a state may result from the two most common scenarios:

I. This is typical of pilgrimage centers of lesser range, where the the number of visitors throughout the year or the pilgrimage season is small and irregular. The emergence of even a small sanctuary service zone requires systematic pilgrimage movement and a relatively constant number of visitors throughout the year. The lack of a sanctuary service zone in smaller pilgrimage centers is associated with the short stay of pilgrims in them, and, certainly, has economic consequences. For many pilgrimage centers of regional range and almost all centers with a smaller range, the lack of the discussed sanctuary service zone is typical, or there is only a single facility there. In the latter case, it will be the initial sanctuary service zone.

1. Sanctuary service zones do not develop in the vicinity of large pilgrimage center complexes with a very extensive and varied infrastructure, as was discussed earlier.

Depending on the way the space around the pilgrimage center and the functions accumulating in sanctuary service zones are organized, they have the following character (Figure 3):

A. Initial—only with individual facilities for pilgrims;

B. Mixed (dispersed);

C. Compact, specialized only in serving pilgrims, with the following subtypes that can be distinguished here: C1—compact linear zones; C2—pilgrimage districts.

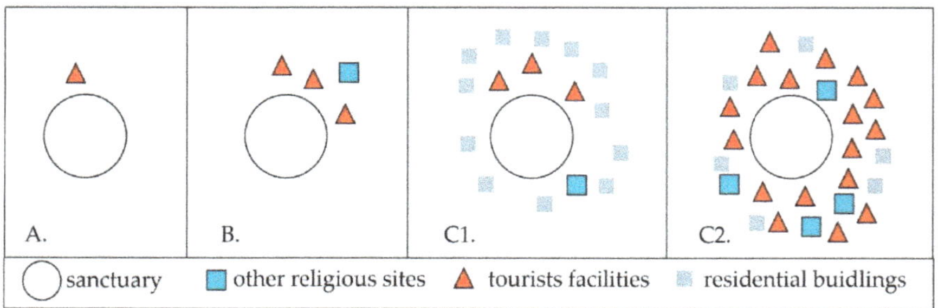

Figure 3. Types of sanctuary service zones: (**A**) initial; (**B**) mixed (dispersed); (**C**) compact: (**C1**) linear zones and (**C2**) districts.

In initial sanctuary service zones (Figure 3A), there is only a single service facility related to the pilgrimage center, typically with a smaller range, especially if these facilities are not present within the pilgrimage center itself.

In mixed (dispersed) zones (Figure 3B), the facilities characteristic of the zone are integrated into the urban development, and functions related to the service of pilgrims coexist with other urban functions. In the landscape of such zones, facilities serving pilgrims are intertwined with facilities for other purposes, most often residential or public buildings, e.g., municipal institutions or educational facilities. The buildings that exist here are usually of dual purpose, with their lower floors occupied by commercial premises, i.e., shops with devotional items, restaurants, and cafés, with the upper floors intended for residential purposes. The mixed sanctuary service zone is the most common zone in large pilgrimage centers. It is characteristic of cities where the pilgrimage center was or is located in the city center, and the spatial development of the city was closely related to it. The city developed around the sanctuary along with (and often much later) pilgrimage infrastructure.

Compact sanctuary service zones (Figure 3(C1,C2)) are observed in the two most common cases: when they are small and cover the nearest area adjacent to the pilgrimage center, most often a street (or even a part of it) or a square, or when they occupy such a large area that they constitute a separate district. A small, compact sanctuary service zone was created, among others, in Kalwaria Zebrzydowska and in Syracuse. A similar zone exists in Cracow-Łagiewniki, which is now starting to transform into a mixed zone. Fatima and Lourdes are typical pilgrimage districts among the surveyed contemporary sanctuaries. A similar district is increasingly visible in San Giovanni Rotondo. Hotel buildings of various sizes and standards are a characteristic type of development in pilgrimage districts. The mentioned San Giovanni Rotondo (especially the part around the sanctuary) has remained a great construction site since Padre Pio's beatification in 1999 and canonization in 2002. The first hotel in Fatima was opened in 1929, and now there are over 50 of them, almost 40 of which are in the pilgrimage district (Soljan 2012). While new hotels are still being built in San Giovanni Rotondo and Fatima, the hotel base in Lourdes has been stable in recent years, especially in the area near the pilgrimage center.

The hotel infrastructure in the zones in question is complemented by other accommodation facilities belonging mainly to church entities (to sanctuary, religious orders, or associations). Fatima is the leader in this aspect, with accommodation services provided by many monastic houses. A well-developed accommodation base can also be found in mixed zones (e.g., Czestochowa and Loreto). The church facilities are a big competition for other facilities.

Apart from the actual areas of sanctuaries, sanctuary service zones are the most significant example of the organization of urban space determined by the pilgrimage function. This is due both to the concentration of facilities and services for pilgrims, unprecedented in other parts of the city, and to the exceptional volume of pilgrimage movement there.

The religious nature of the zones is emphasized by the names of the streets or facilities that are typically religious or otherwise related to the sanctuary.

A noticeable change taking place today in the zones is the appearance of hotel facilities of an increasingly higher standard (e.g., in Lourdes, Fatima, San Giovanni Rotondo, and Czestochowa). At the same time, the number of uncategorized and lowest standard hotels (one- and two-star hotels) is decreasing.

All sanctuary service zones are fairly uniform in terms of the services provided for pilgrims, primarily those related to accommodation and catering. They may be supplemented by photo points, post offices, banks, and ATMs. Additionally, commercial facilities have a very narrow profile of activity. Apart from the numerous shops with devotional items with a fairly wide assortment, there are also jewelry shops with wares traditional for a given region, and bookshops. General grocery shops and chemist's shops are rare.

The infrastructure for the sick and the disabled is not sufficiently developed in all sanctuary service zones (Lourdes is an exception).

Apart from the size of the pilgrimage movement, environmental conditions are a decisive factor in the development and shape of sanctuary service zones. A pilgrimage district may develop, if there is large undeveloped space around a sanctuary, as was the case in Lourdes, Fatima, San Giovanni Rotonto, and the two analyzed sanctuaries in Cracow.

The presented zones are of a conceptual nature. In fact, there are zones with a much more complex structure and difficult-to-define boundaries. This is especially the case when the pilgrimage center itself has open borders—for example, in the case of Altötting, where there is a clear overlap between the sanctuary and the sanctuary service zone. It is worth noting that even among the international and national centers selected for the analysis, there are two, Paris and Levoča, in which a sanctuary service zone has not developed at all. An atypical sanctuary service zone may also be created when a sanctuary is located at a short distance from an object of high tourist attractiveness (in Syracuse and Cracow). This mainly applies to cities with a developed tourist function. This zone serves both pilgrims and tourists, so it partially loses the features typical of a sanctuary zone.

4. Conclusions

Bearing in mind the importance and popularity of religious tourism in the world and its significant impact on, among others, the transformations of geographic space and the emphasized, and still present, gap in the state of knowledge, this article discussed the influence of religious tourism on city space based on the results of research on the development of the sanctuary service zones of the largest European pilgrimage centers. The multifaceted impact of the pilgrimage center on city space can be considered at three scales: macro, meso and micro. This paper is summarized through the proposed conceptual model approach of sanctuary service zone types.

Pilgrimage centers are often a significant factor in urban space. It is enough to mention here, among numerous examples, Lourdes, Fatima, or San Giovanni Rotondo, in which the development of pilgrimage movement has led to major spatial and functional changes in the city.

From the moment of development of pilgrimages, sanctuary service zones may develop in the vicinity of pilgrimage centers, aimed mainly at satisfying tourist needs, in the space outside the sanctuary when not directly inside.

The formation of a sanctuary service zone depends on a number of factors—the most important among them is the range of a pilgrimage center. In most pilgrimage centers with local and regional pilgrimage movement, there are no sanctuary service zones, as there is no demand for tourist infrastructure services. On the other hand, sanctuary service zones are present at large international pilgrimage centers and there are many facilities aimed at satisfying the tourist and material needs of visitors. Pilgrimage traditions also affect the existence and development of sanctuary service zones. For the presence of tourist infrastructure to be justified, pilgrims should stay in a pilgrimage center for more than one day. Historical and socio-cultural factors are also important. Until the occurrence of the socio-cultural changes in the 20th and 21st centuries discussed in the introduction, significantly affecting pilgrimage centers and religious tourism worldwide (Sołjan 2012; Sołjan and Liro 2020), a sanctuary as a sacred space was reserved only for religious functions, and the pilgrims' stay for religious activities. Hence, for centuries, no facilities with non-religious functions (accommodation, catering, and trade) were located inside pilgrimage centers, which was conducive to development of sanctuary service zones. The development of sanctuary service zones is a complex process and also largely depends on other conditions. These include the cooperation of the pilgrimage center with local authorities or even the possibility of expanding and investing in new infrastructure. The development of sanctuary service zones also depends on the size of the city. Research confirmed that the greatest impact is on small- and medium-sized cities. In large cities, especially metropolises, even pilgrimage centers of international range may not cause significant changes, or they are usually limited to the district in which they are located. It is worth paying attention to the fact that, on the other hand, the development of sanctuary service zones may be weakened by investments in tourist infrastructure inside the pilgrimage center. From the second half of the 20th century and at the beginning of the 21st century, owing to socio-cultural changes (Sołjan and Liro 2020), the approach to the idea of sacred space and, consequently, to the location of tourist facilities inside pilgrimage centers changed. Contemporary emerging and developing pilgrimage centers often constitute large sacred complexes in themselves, fulfilling complementary services for visitors, both religious and non-religious. Thus, investments in tourist infrastructure inside pilgrimage centers may limit the development of sanctuary service zones.

Research confirmed that most international and national pilgrimage centers develop sanctuary service zones. The most common are mixed zones, where pilgrimage services coexist with other urban functions. They developed primarily in cities where the pilgrimage center was the main cause of the establishment of a settlement unit, or development of the city followed development of pilgrimage movement and pilgrimage centers. Pilgrimage districts geared towards serving pilgrims developed around the largest pilgrimage centers and, along with the sanctuary, they are often clearly separated from other parts of the city. Linear zones, usually organized along the main street leading to the pilgrimage center, have the least impact and range. Linear zones most often develop in the vicinity of sanctuaries with smaller pilgrimage movement. It is worth noting that to an extent, the influence of the pilgrimage center on the organization of urban space around it decreases with increasing distance from it.

For the proper functioning of a pilgrimage center, it is necessary to manage tourist movement and spatial development within it, as well as in the area of its greatest impact, i.e., sanctuary service zones. Appropriate planning of these spaces may result in strengthening the positive experiences of visitors by comprehensively satisfying their travel needs, extending and diversifying visitor activity zones, extending the duration of stay (Ambrosio 2007; Ambrosio and Pereira 2007; Liro 2021; Liro et al. 2018; Sołjan and Liro 2020), as well as, in a broader sense, increasing the pilgrimage movement and furthering pilgrimage center development, taking into account the multiplier effects, ensuring better economic benefits for the sanctuary and the city (Sołjan 2012). It is worth noting that marketing activities related to the management of a pilgrimage center do not have to contradict the spiritual identity of the place (Shackley 2001). On the other hand, the lack of such activities may cause unplanned and uncontrolled development of services in the sanctuary service zone, adversely affecting the perception of the pilgrimage center (Digance 2003).

Furthermore, sanctuaries also create specific public spaces in cities. Numerous pilgrimage centers are religious centers and, at the same time, places of great tourist attractiveness. In small cities, they are often the main feature attracting visitors with various motivations. The importance of a pilgrimage centers often determines the worldwide popularity and prestige of a city (Lourdes, Fatima, San Giovanni Rotondo, and Czestochowa), affecting the promotion and image of the city, and pilgrimage centers themselves constitute a unique potential for the development of urban space.

Author Contributions: I.S. and J.L. contributed equally to this work. All authors have read and agreed to the published version of the manuscript.

Funding: This research was funded by Institute of Geography and Spatial Management of the Jagiellonian University.

Conflicts of Interest: The authors declare no conflict of interest.

References

Ambrosio, Vitor. 2007. Marketing and development in four western European sanctuary-towns. *International Journal of Management Cases* 93. [CrossRef]

Ambrosio, Vitor, and Margarida Pereira. 2007. Case study 2: Christian/Catholic pilgrimage—Studies and analyses. In *Religious Tourism and Pilgrimage Festivals Management: An International Perspective*. Edited by Razaq Raj and Nigel D. Morpeth. Oxfordshire: CABI, pp. 140–52.

Badone, Ellen, and Sharon R. Roseman. 2004. *Intersecting Journeys. The Anthropology of Pilgrimage and Tourism*. London and New York: University of Illinois Press.

Collins-Kreiner, Noga. 2010a. Researching Pilgrimage. Continuity and Transformations. *Annals of Tourism Research* 37: 440–56. [CrossRef]

Collins-Kreiner, Noga. 2010b. The geography of pilgrimage and tourism. Transformations and implications for applied geography. *Applied Geography* 30: 153–64. [CrossRef]

Collins-Kreiner, Noga. 2019. Pilgrimage tourism-past, present and future rejuvenation: A perspective article. *Tourism Review* 75: 145–48. [CrossRef]

Deffontaines, Pierre. 1948. Géographie et religions. In *Géographie Humaine 21*. Paris: Gallimard.

Digance, Justine. 2003. Pilgrimage at contested site. *Annals of Tourism Research* 301: 143–59. [CrossRef]

Eade, John. 2017. Parish and Pilgrimage in a Changing Europe. In *Migration, Transnationalism and Catholicism. Global Perspectives*. Edited by Dominic Pasura and Marta Bivand Erdal. Basingstoke: Palgrave Macmillan, pp. 75–92.

Eade, John, and Michael Sallnow, eds. 1991. *Contesting the Sacred. The Anthropology of Christian Pilgrimage*. London and New York: Routledge.

Fickeler, Paul. 1947. Grundfragen der Religionsgeographie. *Erdkunde* 1: 121–44. [CrossRef]

Giovine, Michael A. 2015. When Popular Religion Becomes Elite Heritage: Tensions and Transformations at the Shrine of St. Padre Pio of Pietrelcina. In *Encounters with Popular Pasts. Cultural Heritage and Popular Culture*. Edited by Mike Robinson and Helaine Silverman. Cham: Springer International Publishing, pp. 31–47.

Gökariksel, Banu. 2009. Beyond the officially sacred. Religion, secularism, and the body in the production of subjectivity. *Social & Cultural Geography* 10: 657–74. [CrossRef]

Holloway, Julian, and Olivier Vallins. 2002. Placing religion and spirituality in geography. *Social & Cultural Geography* 3: 5–9. [CrossRef]

Kaufman, Susanne. 2005. *Consuming Visions. Mass Culture and the Lourdes Shrine*. Ithaca and London: Cornell University Press.

Liro, Justyna. 2018. Organizacja Przestrzeni i Funkcje Sanktuariów na Przykładzie Wybranych Ośrodków Pielgrzymkowych w Polsce [Organization of Space and Functions of Sanctuarias on the Example of Selected Pilgrimage Centres in Poland]. Ph.D. thesis, Institute of Geography and Spatial Management, Jagiellonian University, Cracow, Poland.

Liro, Justyna. 2021. Visitors' motivations and behaviours at pilgrimage centres: Push and pull perspectives. *Journal of Heritage Tourism* 16: 79–99. [CrossRef]

Liro, Justyna, and Izabela Sołjan. 2016. The Development of the Sacred Space and Pilgrimage Movement in the Context of Identity Formation—Example of Sanctuary of Our Lady of Lichen. In *Pilgern, Wege, Heilige Orte. Seelsorge im Kontext von Menschen Unterwegs*. Edited by Michaela C. Hastetter and Maciej Ostrowski. Cracow: Unum Publishing, pp. 421–27.

Liro, Justyna, Izabela Sołjan, and Elżbieta Bilska-Wodecka. 2017. Visitors' diversified motivations and behavior—The case of the pilgrimage center in Krakow Poland. *Journal of Tourism and Cultural Change* 15: 1–20. [CrossRef]

Liro, Justyna, Izabela Sołjan, and Elżbieta Bilska-Wodecka. 2018. Spatial Changes of Pilgrimage Centers in Pilgrimage Studies—Review and Contribution to Future Research. *International Journal of Religious Tourism and Pilgrimage* 6: 5–17. [CrossRef]

Nolan, Mary L., and Sidney Nolan. 1989. *Christian Pilgrimage in Modern Western Europe*. London: University of North Carolina Press.

Nolan, Mary L., and Sidney Nolan. 1992. Religious sites as tourism attractions in Europe. *Annals of Tourism Research* 191: 68–78. [CrossRef]

Reader, Ian. 2014. *Pilgrimage in the Marketplace*. New York and London: Routledge.

Rinschede, Gisbert. 1987. The Pilgrimage Town of Lourdes. *Journal of Cultural Geography* 7: 21–33. [CrossRef]

Rinschede, Gisbert. 1988. The Pilgrimage Center of Fatima/Portugal. *Geographia Religionum* 4: 65–98.

Rinschede, Gisbert. 1995. The pilgrimage center of Loreto, Italy. *Pilgrimage Studies. Text and Context* 3: 157–178.

Rinschede, Gisbert, and Angelika Sievers. 1987. The pilgrimage phenomenon in sociogeographical research. *The National Geographical Journal of India* 33: 213–17.

Shackley, Myra. 2001. Managing sacred sites: Service provision and visitor experience. *Material Religion the Journal of Objects Art and Belief* 1: 428–29.

Shinde, Kiran, and Daniel Olsen. 2020. *Religious Tourism and the Environment*. CABI Religious Tourism and Pilgrimage Series. Available online: https://www.cabi.org/bookshop/book/9781789241600/ (accessed on 20 January 2021).

Shinde, Kiran, and Katia Rizello. 2014. A Cross-cultural Comparison of Weekend–trips in Religious Tourism: Insights from two cultures, two countries (India and Italy). *International Journal of Religious Tourism and Pilgrimage* 2: 17–34. [CrossRef]

Sołjan, Izabela. 2011. Sanktuarium jako organizator przestrzeni miejskiej (na przykładzie wybranych sanktuariów w Polsce) [The sanctuary as a means of organizing urban space: A case study of selected sanctuaries in Poland]. *Tourism* 21: 49–58. [CrossRef]

Sołjan, Izabela. 2012. *Sanktuaria i ich Rola w Organizacji Przestrzeni Miast na Przykładzie Największych Europejskich Ośrodków Katolickich [Sanctuaries and Their Role in the Organization of Urban Space on the Example of Europe's Biggest Catholic Centres]*. Cracow: Institute of Geography and Spatial Management, Jagiellonian University.

Sołjan, Izabela, and Justyna Liro. 2015. Nowe centrum pielgrzymkowe w krakowskiej dzielnicy Łagiewniki-Borek Fałęcki [A new pilgrimage centre in the Cracow district of Łagiewniki-Borek Fałęcki]. *Peregrinus Cracoviensis* 25: 135–51. [CrossRef]

Sołjan, Izabela, and Justyna Liro. 2020. The Changing Roman Catholic Pilgrimage Centers in Europe in the Context of Contemporary Socio-Cultural Changes. *Social & Cultural Geography*. [CrossRef]

Sopher, David E. 1981. *Geography of Religion*. Englewood Cliffs: Prentice-Hall.

Article

Constructing and Contesting the Shrine: Tourist Performances at Seimei Shrine, Kyoto

Mia Tillonen

Graduate School of International Media, Communication, and Tourism Studies, Hokkaido University, Sapporo 060-0817, Japan; mia.tillonen@gmail.com

Abstract: Japanese Shinto shrines are popular pilgrimage sites not only for religious reasons, but also because of their connections to popular culture. This study discusses how tourism is involved in the construction of the shrine space by focusing on the material environment of the shrine, visitor performances, and how the shrine is contested by different actors. The subject of the study, Seimei Shrine, is a shrine dedicated to the legendary figure Abe no Seimei (921–1005), who is frequently featured in popular culture. Originally a local shrine, Seimei Shrine became a tourist attraction for fans of the novel series Onmyōji (1986–) and the movie adaptation (2001). Since then, the shrine has branded itself by placing themed statues, which realize the legend of Abe no Seimei in material form, while also attracting religious and touristic practices. On the other hand, visitors also bring new meanings to the shrine and its objects. They understand the shrine through different kinds of interactions with the objects, through performances such as touching and remembering. However, the material objects, their interpretation and performances are also an arena of conflict and contestation, as different actors become involved through tourism. This case study shows how religion and tourism are intertwined in the late-modern consumer society, which affects both the ways in which the shrine presents and reinvents itself, as well as how visitors understand and perform within the shrine.

Keywords: shinto shrine; popular culture; tourist performances; material religion

Citation: Tillonen, Mia. 2021. Constructing and Contesting the Shrine: Tourist Performances at Seimei Shrine, Kyoto. *Religions* 12: 19. https://dx.doi.org/10.3390/rel12010019

Received: 20 November 2020
Accepted: 21 December 2020
Published: 28 December 2020

1. Introduction

In Japan, Shinto shrines are popular destinations for both domestic and inbound tourists[1]. Not only the traditionally famous shrines, but also smaller, more local shrines have become popular through tourism. Travel to religious places is not a new phenomenon, as Japan has a long-lasting tradition of pilgrimage (Nakanishi 2018). However, as society changes, so does the role and form of religion and travel. Some forms of religious tourism[2], such as *hatsumōde*, the first visit to the shrine after New Year's, and pilgrimage tourism, such as bus tours in Shikoku island, were born through the involvement of the tourism industry and the development of transportation systems (Hirayama 2015; Kadota 2013). Nowadays, Shinto shrines are often visited because of images created by mass media. Since the 2010s, a nation-wide power spot boom[3] and the collection of seal stamps (*goshuin*) have made visiting shrines a popular past-time (Suga 2010; Okamoto 2019a). In addition,

[1] For example, in a Kyoto tourism survey, shrines and temples were given as the most popular reason for traveling to Kyoto for both Japanese as well as foreign visitors (Industry and Tourism Bureau 2019).

[2] In Japanese, tourism in religious places is called *shūkyō tsūrizumu* (literally, religion tourism), which is not the exact equivalent of religious tourism. The term itself does not take a stance on whether the tourist is traveling for religious reasons or not, whereas the English term includes religious motivation behind travel. This can be seen as a major difference in the focus of the field in Japan.

[3] The term power spot refers to a location in which one can feel a strong, invisible spiritual power, energy or *ki* (Japanese for *qi*) (Horie 2017, p. 192). While the idea of power spots emerged in the mid-1980s via the global New Age movement (Carter 2018), it entered the vocabulary of mainstream Japanese society in the 2010s through media representations, such as women's magazines (Tsukada and Omi 2011). Many pre-existing religious places, such as Shinto shrines, have been reframed as power spots by the mass media (Okamoto 2019a).

as shrines are frequently featured in popular culture, many fans want to visit the places connected to their favorite movies or games (Yamamura 2015). This is why the Japanese word for pilgrimage (*seichi junrei*) has received new connotations within anime and games—not just religion. Thus, it can be said that shrines have become tourist attractions for people with various motivations, which combine both secular and religious meanings.

In recent years, the study of religion and tourism, as well as modern pilgrimage, has become a growing field worldwide (Collins-Kreiner 2020), and also in Japan, where the touristic use of religious practices and places has been discussed by scholars of religious studies and tourism studies since the 2010s. In fact, scholars of religious studies have taken up the topic of tourism, not only as proof of the secularization of religious traditions but as a legitimate object of research that reflects the state of modern religion (Suzuki 2020). In previous studies, tourism in religious places has been described as the consumption of representations (Yamanaka 2012, 2015), where religious places and practices are given new value and meaning, as 'something worth visiting' not just in the religious sense. In particular, mass media has a great influence in the process of recontextualizing traditionally religious places and emphasizing the spiritual side, making them easier to add to travelers' itineraries (Okamoto 2015). Using various strategies, such as branding and collaborating with well-known characters and celebrities, religious organizations in contemporary Japan are using, reshaping, and updating their religious sites and traditions in order to compete in a secularized world (Porcu 2013, 2014). It can even be argued that market engagement is crucial for the continuation of religious practices such as pilgrimage (Reader 2014). Moreover, the consumption of religion outside of its conventional context has been seen to reflect the emergence of a new spiritual market, where the line between the religious and the secular is fluid. Within the spiritual market, not just religious institutions, but also actors from the cultural industry (including tourism) provide products such as religious tourism to satisfy the consumers' spiritual needs (Yamanaka 2016, 2017, 2020).

The touristification of sacred places has also been emplaced within a wider process of sacralization, in which the reinvention of shrines as power spots is not driven by religious institutions themselves, but rather by outside actors (Rots 2017). This movement, however, is paralleled with attempts to de-privatize Shinto and to redefine and reconfigure beliefs, practices and institutions previously classified as 'religion' as Japanese traditional culture and heritage (Rots 2017, p. 190). On the other hand, the involvement of outside actors has also been problematized when it comes to using religion as a tourism resource. For example, as the government is not legally authorized to promote religious rituals, in the promotional materials of the Gion matsuri in Kyoto, the 'cultural' side has been emphasized as a strategic move to gain UNESCO recognition (Porcu 2012). In addition, Foster (2020) discusses how the Namahage tradition of Akita prefecture has been stripped of its religious characteristics when being rebranded as intangible cultural heritage. However, in the tourism-oriented Namahage Sedo Matsuri, the religious dimensions of the practice have been highlighted, even more than the actual New Year's Eve ritual of the locals. Kadota (2013) argues that when designated as heritage, the concept of value is brought to religion, as the meaning of religious places, things, and practices are evaluated from the outside. As certain religious practices are included, whereas others are excluded, criteria of so called 'desirable' and 'undesirable' faiths are created (Kadota 2013, p. 73). Thus, the same tradition is treated, practiced and used differently depending on the context and the actors involved.

The perspectives discussed above can be used as a way to explain *why* Shinto shrines become tourist attractions and how secular actors, such as local governments, UNESCO and the tourism industry, have become involved with their different policies and agendas. However, these approaches focus mostly on the macro level and tend to dismiss the material and corporeal side of tourism and religion in favor of the representational. Therefore, in this paper, I will focus on the material aspect of religion and tourism in order to analyze the *how* of tourism in Shinto shrines. Through the case study of Seimei Shrine, Kyoto, I will explore how shrines become tourist attractions: how do visitors perform and understand the shrine? How is tourism emplaced and enacted materially within the shrine? What kind of conflicts occur in this process?

The article is structured as follows. The first part introduces the process through which different media representations have changed Seimei Shrine from a local shrine to a tourist attraction. The following part will then analyze how this has affected the construction of the shrine space by focusing on the material environment, visitor performances, and how the shrine is contested by different actors. The last part emplaces the case study within a wider discussion about religious change in the late-modern consumer society.

2. Research Methods

This paper focuses on tourism in Seimei Shrine observed in the material environment and different visitor performances. It is based on qualitative data gathered through two periods of fieldwork in January and September 2018, during which the author paid regular visits to the shrine and conducted interviews with its visitors within opening hours. Newspaper articles, the official website, and social media accounts of Seimei Shrine and TripAdvisor comments were used as secondary data.

Participant observation was conducted within the shrine premises through photography and fieldnotes of how visitors interacted with the objects. Semi-structured interviews were conducted with one hundred visitors in front of the shrine. The interviewees were not participants in tours, but individual travelers, either traveling alone or with a partner or a family member, mostly from outside of Kyoto. Most of them were in their 20s and 30s. The interviewees were asked various questions, including the reason for their visit, their previous image of the shrine and their experience at the shrine. The interviewees were also asked to show the route of their visit, in which order they interacted with different objects. Through content analysis, the interview data were divided into themes, such as the reason for visiting, one's prior image of the shrine, purchases and interactions with objects. These themes were used as the basis for the analysis in Part 4.

As Nelson has pointed out, it is not unproblematic to make definitive statements about the motivations behind an individual's visit to a shrine only by watching or conducting interviews (Nelson 2000, p. 27). This is why it is important to acknowledge the limitations of the interview data and especially the relative shortness of individual interviews (spanning from 5 to 15 min). However, due to the involvement of the tourism industry and the popularity of power spots, similar changes and performances can be observed in other shrines as well. Therefore, this current study can be used as a reference for further research.

3. The Case of Seimei Shrine, Kyoto

First, let us take a look at how Seimei Shrine has become a tourist attraction, and how it has been represented in the media. Seimei Shrine is a relatively small shrine located in the Kamigyō ward of Kyoto. It is a shrine dedicated to Abe no Seimei (921–1005), an *onmyōji* [4] of the Imperial office during the Heian period (794–1185). As a local shrine, Seimei Shrine itself is known for its benefits of preventing danger (*yakuyoke*). Every year the shrine brings residents and other people involved together for different festivals, such as the Seimei Festival (*Seimei-sai*) in September, during which the shrine is decorated with lanterns and events such as a parade are organized[5]. The presence of the shrine can also be observed in

[4] *Onmyōji* is a practitioner of Onmyōdō, which can be translated as 'The Way of Yin-yang'. In the Heian period, *onmyōji* were civil servants in the Onmyōdō bureau in the *ritsuryō* system of governance. They were experts in astrology, calendar studies and divination. There are debates whether Onmyōdō should be considered a religion or a technique (see Hayashi and Hayek 2013).

[5] On recent discussion about tourism and festivals (*matsuri*), see the Special Issue of *Journal of Religion in Japan* Vol. 9 (2020).

the surrounding area, as many residents and local businesses place paper talismans (*ofuda*) purchased at the shrine close to their entrance[6].

While Seimei Shrine was established a thousand years ago, its career as a tourist attraction started in the early 2000s. Originally a mystical figure, Abe no Seimei's legends have been preserved in traditional tales[7]. In the early 1990s, he became a household figure in popular culture, when he was featured in Baku Yumemakura's novel series Onmyōji (1986–) and later Reiko Okano's manga series based on the novel (1993–). However, the so-called Seimei boom reached its peak when *Onmyōji*, an NHK television drama, starring Goro Inagaki, a member of the boyband SMAP, and a movie[8] starring the actor Mansai Nomura, were released in 2001. These depictions turned Abe no Seimei from a 'grave, middle-aged man exemplary of Heian-era masculinity' to a beautiful young man (*bishōnen*), suitable for the aesthetic tastes and desires of young women (Miller 2008, pp. 31–32). This 'extreme makeover' led to young female visitors visiting Seimei Shrine and writing their messages to Abe no Seimei on wooden votive tablets (*ema*).

In the late 2000s and early 2010s, overlapping with an increase in visitor numbers to Kyoto, a national power spot boom occurred. This time, Seimei Shrine's sacred tree (*goshinboku*) caught the visitors' interest, as it was said to have special spiritual energy. The tree was also featured in articles covering the boom and a picture of visitors touching the tree was published in the Asahi Shimbun newspaper (Asahi Shimbun 2010). Fitting the image, the shrine placed a sign encouraging touching the bark of the tree as well as the construction of a platform to protect the tree's roots from the damage caused by the increasing number of visitors.

Recently, Seimei Shrine has also been linked to the figure skater Yuzuru Hanyu, who won an Olympic gold medal for performing to music of the movie *Onmyōji* in 2018. Hanyu's visits to the shrine have been covered by the media and have led to Seimei Shrine becoming a pilgrimage site for the skater's fans. Sports magazine Nikkan Sports, for example, writes about Hanyu's visit as follows[9]:

> On the 2nd of July, [Hanyu] visited Seimei Shrine in order to get a deeper understanding [of Abe no Seimei]. Immediately after entering the shrine premises, he showed wariness and said 'I can feel *ki* [energy]'. He suddenly realized that he will perform the actual presence of a god. He was shown the portrait of Abe no Seimei from the shrine's private collection. Also by placing his hands on it, he received power from the 300 year old sacred tree. Almost as if he was asking Abe no Seimei for permission [to perform], he walked around the shrine premises for an hour, although it would normally take only ten minutes.
>
> (Takaba 2015)

This kind of coverage resulted in Seimei Shrine becoming a pilgrimage site for the skater's fans. The same way as before, this can be seen in the visitors' personalized votive tablets dedicated to the skater's success. Seimei Shrine has also embraced the attention both virtually, as the main priest of the shrine congratulated Hanyu's success on Seimei Shrine's

---

[6] This paper focuses on visitors from outside of the shrine. However, the special relationship between the local community and the shrine should be noted, as many neighborhoods are closely linked to the area's Shinto shrine. As the local deity (*ujigami*) of the neighborhood is revered in the shrine, the 'group from the land surrounding the areas dedicated to the belief in and worship of one shrine; or, the constituents of that group' are traditionally referred to as *ujiko* (Sano 2007). For more on *ujiko* and the role they play in for example local festivals, see (Porcu 2012; Sonoda 1975). In addition, the term *sūkeisha* is used, but 'the two are distinguished by a geographical classification with *ujiko* referring to the person from that shrine's *ujiko* district and *sūkeisha* referring to the person from outside the district' (Sano 2007). Ishii has noted that the role of *sūkeisha* will become greater due to the weakening of the consciousness of belonging to one's local shrine (Ishii 2010). In fact, Seimei Shrine established an organization (*sūkeikai*) for patrons outside of the local community to which anyone can participate by paying a membership fee

[7] In popular culture, Abe no Seimei has become a sorcerer-type hero figure, who protects the Heian court from evil spirits and other enemies. For details of Abe no Seimei's life based on historical records, see for example Shigeta (2013). For more on the so-called Seimei boom, see Miller (2008). This paper will not focus on the worship of Abe no Seimei or the practice of Onmyōdō, but the process through which the shrine has become a tourist attraction.

[8] The movie was released outside Japan under the title of *Onmyoji: The Yin Yang Master* (2001). With a box-office gross of approximately 3 billion yen, the movie was the fourth most popular movie of the year 2001 (Motion Picture Producers Association of Japan Inc 2001).

[9] All the translations are by the author.

official Twitter site, as well as physically at the shrine with congratulatory messages paired with the selling of an amulet for winning (*katsu mamori*).

Through this brief overview, we can see how Seimei Shrine turned from a local shrine to a tourist attraction due to different media representations and fan performances. As Shinde (2013) pointed out, religious narratives and the various media they are presented in occupy an important position in the service of religious tourism, as they also reflect contemporary religious tastes. From one boom to another, Seimei Shrine has been reframed several times by the mass media, making it a pilgrimage site for several fandoms. While the impetus for tourism originally came from outside, Seimei Shrine also started actively changing its own image and material environment through branding in the 2000s. This is why focusing solely on media representations does not give a full picture of the process through which Seimei Shrine has become a tourist attraction.

4. Constructing, Performing and Contesting Seimei Shrine

How is the shrine space produced and how do we understand tourism within that process? When discussing the poetics and politics of American sacred spaces, Chidester and Linenthal (1995) introduce three features through which sacredness is produced: (material) construction, rituals, and contestation by different actors. The production and contestation of tourist space has also been the focus of the performative turn in tourism studies (Bærenholdt et al. 2004; Edensor 1998, 2001; Haldrup and Larsen 2010; Urry and Larsen 2011). The performative turn has emphasized the corporeal and material aspect of tourism—opposed to previous representational approaches which privileged the visual experiences of tourists, such as the notion of 'the tourist gaze' (Urry 1990). Tourist places are not just seen as passive stages for performances, but as produced places, which tourists co-produce through their performances (Bærenholdt et al. 2004, p. 10). The performative turn thus draws attention to how tourists experience and construct places in multi-sensuous ways, which include different interactions with the material world (Haldrup and Larsen 2010).

In this article, special focus will be given to the role of material objects and the practices they are involved in at Seimei Shrine. This is aligned with the interests of the study of material religion, a perspective on religious studies, which has gathered popularity since the 2000s (Hirschkind 2006; Keane 2007; Meyer 2009; Morgan 2010; Vásquez 2011). Within the study of material religion, the focus has shifted from studying texts and meaning to the material forms and embodied practices of religion. Through this shift, the role of material objects has changed from vessels of underlying meaning to actors within various contexts and practices. To quote the editors of the journal *Material Religion*, 'the meaning of an object is not understood to reside singularly inside in it, but to also draw from its circulation, its local adaptation, from what people do with it, and from the affective and conceptual ways whereby users apprehend an object' (Meyer et al. 2010, p. 209). In addition, in the context of Japan, the study of materiality has been gathering interest among scholars[10]. For example, Daniels has explored objects of good luck (*engimono*) and spirituality in the domestic environment (Daniels 2003, 2012). In addition, Rambelli's work has showed how materiality is an intrinsic part of Japanese Buddhism, not only in popular practices but also in scholastic and doctrinal themes (Rambelli 2007).

However, in tourism studies, the discussion of the material side of religious tourism is quite recent. Focusing on materiality allows us to grasp the improvisational side of religious tourism, for example the different understandings of religious places and experiences (Terzidou 2020). Through this we can understand that visitors' actions are not confined to symbolic meanings but also the shapes and sizes, surfaces and locations of objects. However, religious sites and experiences are not confined to fixed places, as the portability of material objects enhances a sense of connectedness and *communitas* even for those who are not physically present (Higgins and Hamilton 2020; King 2010). Therefore, instead of separating the representational and material aspect, the relationship between religion and

10 For example, a Special Issue of the *Japanese Journal of Religious Studies* 45/2 (2018) was dedicated on the topic.

tourism should be understood as a form of experiencing religious sites through interaction between people, objects, and the environment.

Using Chidester and Linenthal's (1995) three aspects as a framework, in the following analysis I will trace the role material objects take as the shrine is constructed, performed, and contested by various actors. Through this, I aim to shed light on the different ways in which religion and tourism are entwined within the shrine and the tourists' performances and interactions with material objects.

4.1. The Construction of Seimei Shrine—Theming the Shrine

Since the mid-2000s, Seimei Shrine has established itself as a tourist attraction in Kyoto. In order to adapt to the growing number of tourists, the shrine has not only improved its infrastructure with multiple parking areas and a new bus stop, but also implemented strategic branding and image management (Einstein 2007; Porcu 2014). This can be observed both in the virtual presence (official website, social media), as well as the physical environment of the shrine. The most apparent aspects are theming and merchandising (Bryman 2004), as the theme of Abe no Seimei and his legends are present in the overall visual look of the shrine, from their website and pamphlets, to bronze statues as well as other paraphernalia that can be purchased at the shrine.

Let us start with the material construction of the shrine. Seimei Shrine is divided to two areas with torii gates, the outer and inner part. The inner part, which contains the main altar (*honden*) and the shrine office, hosts objects which are common to most of the shrines in Japan. An example of this is the *temizusha*, where one can ritually purify oneself before entering the shrine premises. While being the religious part of the shrine, the inner part also has objects that cannot be found anywhere else other than Seimei Shrine. These include objects, such as a bronze statue of a peach, which is said to ward off evil, a statue of Abe no Seimei, and a pentagram-shaped well[11]. There are also more touristic objects, such as picture boards, which explain the different legends, as well as a cardboard cutout of Abe no Seimei, which visitors can use to take pictures[12]. The outer part has the official souvenir shop Kikyōan and various themed stone statues, such as the replica of the famous Ichijō bridge and beside it a statue of a *shikigami*, a gremlin-like character, who is also featured in the legends.

The various statues are explained to have special religious meaning and spiritual power. The water of the pentagram-shaped well, for example, is said to have healing properties. On the other hand, the same objects are also given distinct stories—it is said that Abe no Seimei used his magical powers to make water burst out of the ground, which was then where the well was built. These stories and meanings are conveyed through signs, which are placed next to the objects to explain their meaning to the visitors. Despite the fact that the statues have been built recently, the signs emphasize the objects' authenticity by contextualizing them as history and folklore. For example, the bronze statue of the peach is explained as follows:

> Since ancient times, in China as well as in the practice of Onmyōdō, the peach has been told to be a fruit that protects against misfortune. Even in Kojiki and Nihonshoki, the peach has been described as warding off evil. This can be seen as the origin of the folktale Momotarō, too. Everyone has their own bad luck and misfortunes. By stroking away the bad luck onto this peach, you can feel refreshed.
>
> (Sign explaining the statue)

[11] Onmyōdō is a Japanese esoteric cosmology, which is based on ancient Chinese philosophy. The pentagram (*gobōsei*) is said to symbolize the five elements (*gogyō*), which are the basic principles of Onmyōdō along with yin-yang (*onmyō*). It is also used as the logo of Seimei Shrine and can be found in different forms around the shrine.

[12] Interestingly, while the statue is made based on a Heian-era portrait of a middle-aged Abe no Seimei, the illustrations on the picture boards feature Abe no Seimei as a handsome young man, adapting to the current image of the wizard.

For most visitors, the different signs act as the main source of information. They are 'on-site markers' (MacCannell 1999), which also affect the visitors' interpretations and behavior, showing them what is 'worth seeing' at the shrine. The explanations use different narrative elements in order to convey the official interpretations of the objects as well as the general theme of the shrine. Connecting the peach statue to famous folktales enhances the story-like theme of the shrine, whereas by encouraging physical engagement, the peach is given a certain spiritual status as an object that wards off evil, thus combining mythical elements with a concrete suggestion of action (Figure 1). The sign also has a commercial function, as below the explanation the shrine promotes a special peach amulet sold at the shrine office.

Figure 1. Visitor touching the peach statue while reading the sign.

While the statues were built in the 2000s, the sacred tree has been growing within the premises since long before the tourists arrived. However, the touristification of the shrine and the nationwide power spot boom made it into a popular attraction among the visitors. In fact, the tree has been reframed as a power spot (both by the mass media and the shrine itself), and is frequently featured in guidebooks and magazines. Most visitors engage with the tree by touching it. Seimei Shrine also encourages this kind of behavior: 'The bark of the tree has a unique feel to it. Press both of your hands on it to feel the power of the great tree.' In addition, when asked what was memorable at the shrine, many of the visitors answered the sacred tree. While touching sacred trees is not allowed in all shrines (sometimes even explicitly prohibited, such as in Ise Shrine), it is a common practice at many so-called power spots (Horie 2019). It can be said that this kind of physical engagement is a way through which visitors understand places and objects as power spots. This also brings about certain expectations of a place to which shrines have to adapt.

The nationwide power spot boom has also changed the 'imaginary geography' (Larsen 2006) of Shinto shrines. Much of this can be attributed to the influence of celebrity spiritual counselor Hiroyuki Ehara, who published a series of books that introduced Japanese shrines as power spots (Horie 2019, p. 178). Magazines and television shows feature power spots in mysterious locations, usually surrounded by nature—something one could find in a Hayao Miyazaki movie. This image was also held by the interviewees—while some explained that the shrine is easy to approach due to the fact that it is a power spot, some commented on the shrine being too urban compared to the natural image they had. For example, the interviewees below compared Seimei Shrine to Shimogamo Shrine, another famous shrine in Kyoto:

> Visitor A: Shimogamo Shrine has a lot of green. It's a forest, so I received power [energy] there. I wonder what the difference [with Seimei Shrine] is? It feels more sacred there. It might be because Seimei Shrine has too many visitors? And the trees are all planted.
>
> Visitor B: It feels more sacred, when there's nature.
>
> Visitor A: Of course [Seimei Shrine] is sacred. But because there are a lot of people here, Shimogamo Shrine feels more soothing.
>
> (Visitors in their 30s)

For visitors, the natural environment plays a great role in the construction of the power spot. That is also why the built environment and other people feel distracting. The demand for a soothing atmosphere reflects the image of power spots being places of healing and relaxation—the emphasis is not on the religious but on the spiritual side. They are not places where one would necessarily engage in institutional rituals, for example. Instead, the visitors 'receive energy' and feel refreshed by visiting natural environments and touching various objects—similar performances can be observed in other power spots as well.

Not all shrines are positive towards the power spot trend, as they fear that visitors might neglect the actual place of worship. This represents a gap between what institutions and visitors regard as the object of worship (Okamoto 2019a, p. 136), which at times contests the official interpretation of the shrine[13]. This is not only seen on the level of performances, however; it can also be observed in the material environment of the shrine. To adapt to visitors' needs, religious places have started using various management techniques, such as zoning and limiting entrance (Kadota 2016). Seimei Shrine has limited the flow of visitors so that the inner part of the shrine can be entered during the shrine's opening hours (9:00–16:30). The shrine has also built a platform around the sacred tree to protect its roots from the growing number of visitors wanting to touch it.

Furthermore, it can be argued that since the 2000s, Seimei Shrine has gone through a reflexive process, which can be observed in the increase in various objects. Through the theming of the environment and objects, the shrine has strategically branded itself in order to suit visitors' needs, as well as to distinguish itself from other tourist attractions in Kyoto. While visiting a shrine includes physical engagement with the environment, such as washing one's hands and mouth with water, Seimei Shrine has turned touching certain objects into a special activity. Engaging with the objects, the visitors' actions enhance both the objects' and the shrine's new meanings. On the other hand, the shrine has also used various management techniques in order to control the visitors' actions and interpretations. Thus, by encouraging and limiting certain behavior, the shrine actively produces the experiences of the visitors not just as a religious institution but as a tourist attraction (Paine 2019; Shinde 2020).

4.2. Performing the Shrine—The Different Meanings People Bring to the Shrine

Without people, there would be no shrine. When entering from the main street, the visitors are welcomed by the first torii gate, which has the pentagram symbol on it. After taking pictures of the gate, most of the visitors walk straight to the inner shrine. They purify themselves at the *temizusha* located close to the entrance before paying their respects

[13] Not only between individual Shinto shrines and visitors, this gap can be observed on a national level. For example, the umbrella organization for Shinto shrines, Jinja Honchō, has dismissed the popularity of power spots as a short-term trend that does not support long-term engagement (Carter 2018, p. 162). Carter argues that by delineating power spot related practices, the organization attempts to establish a set of norms for shrine worship which points to an underlying concern regarding Jinja Honchō's authority over ritual and practice (Carter 2018, p. 163).

at the main altar. After that, they engage with the more unconventional objects by stroking the bronze statues and the sacred tree. The visitors also purchase different religious objects, such as amulets. Before leaving the shrine, many take a look at the souvenir store located in the outer part of the shrine.

In this way, the shrine is constructed through different practices. However, it is not just the religious rituals but also the different tourist performances that produce the space (Edensor 1998, 2000). While tourist performances are greatly influenced by different narratives, the performative processes themselves reconstitute the symbolic value of the site. In fact, sites 'tend to be fluid entities whose meanings and usage change over time and are apt to be contested by different tourist groups' (Edensor 2000, p. 326). Thus, different meanings and media representations are emplaced not only by the shrine, but also through the visitors' performances. However, the representational side is only one part of the story, as the performances are bound in a network of material relations between people, objects and the environment.

While Shinto lacks strict doctrines, and practices in shrines are relatively free of restrictions (Nelson 2000), in Seimei Shrine there are certain 'disciplined rituals' (Edensor 2000), common practices based on social etiquette. These actions are produced through socially acquired knowledge, the presence of other visitors, and the engagement with the material environment of the shrine, producing the flow of performances. However, this kind of common knowledge applies only to objects, which can be found in other shrines. Regarding the special objects, such as the themed bronze statues, the visitors tend to partake in more improvisational performances. Many, for example, made funny poses and took pictures with the peach statue, and laughed when talking about it. So-called post-tourist performances (Edensor 1998) could also be observed, especially with the more overtly touristic objects, such as the Abe no Seimei picture board and Instagram panel, with which some visitors took funny poses and commented on sarcastically.

However, while improvisational, the performances are not completely arbitrary, as they 'rely on contexts and instructions to provide a broad framework within which some improvisation may take place' (Edensor 2000, p. 335). In the case of Seimei Shrine, the context also relies heavily on popular culture. As mentioned above, Seimei Shrine became famous through movies and TV series, and therefore is visited by members of several fandoms. Thus, so-called popular remembrance, such as that of different media representations, influences the ways in which people understand and talk about the shrine. It has been said that through the consumption of different media representations, visitors connect to, and enhance their experience of the place (Timm Knudsen and Waade 2010, p. 12). In fact, when asking about where the interviewees first heard about the shrine, while some mentioned the shrine's connection to the movie *Onmyōji*, most talked the figure skater Yuzuru Hanyu. They connect its popularity with the figure skater—not the religious aspect or the deity of the shrine:

> [Seimei Shrine] has become a famous shrine. Compared to before, this is all thanks to Hanyu (laugh). I do not think that people really knew much about *onmyōji* or Abe no Seimei, that is why [it became famous].
>
> (Visitor in their 50s)

The connection to Yuzuru Hanyu can be observed in material objects within the shrine. For example, the shrine displays a wooden tablet written by the figure skater for fans to see, and some of the interviewees specifically came to see his autograph. The shrine also sells a winning amulet connected to the skater's success in the Olympic games. However, Hanyu's presence is most visible in the wooden votive tablets (*ema*), through which fans express their devotion to the skater.

Ema are wooden votive tablets to which one can write their wishes or vows. Usually, visitors purchase and personalize the votive tablets 'on the spot' and leave them at the shrine. They act as a medium through which the wishes or needs are made known to the deity enshrined in a shrine or temple (Reader 1991, p. 23). Moreover, they have also become an intrinsic part of Japanese fan culture. Although often connected to anime pilgrimage, other fandoms express themselves through votive tablets as well. Reader, for example, pointed out that votive tablets 'provide a scope and setting whereby baseball fans can bear witness to their support for their favorite team' (Reader 1991, p. 37). This communicative aspect has been discussed by Andrews, who noted that instead of fans communicating face-to-face, their 'communication is anchored in the material plane through the medium of the *ema*' (Andrews 2014, p. 220).

Who are the messages meant for, then? In anime pilgrimage, it seems to be more towards the fans than the deities worshiped in the shrine, as sometimes they act as a physical forum through which fans communicate with each other (Andrews 2014, p. 221). This can be seen in the illustrations and messages addressed to other *ema*s hanging on the rack. In her work on the Seimei boom among Japanese schoolgirls in the 2000s, Miller also noted that instead of addressing their messages to Abe no Seimei, visitors wrote messages to the novelist Yumemakura Baku and manga artist Reiko Okano, who made the wizard popular in the first place (Miller 2008, p. 37).

In addition, in the case of Yuzuru Hanyu, *emas* connect their writers not only to the shrine but also to the figure skater, as well as the wider fan community. This can be seen as being rooted in, as Reader explains, the humorous and ludic dimensions to *ema*, as well as their accessibility and the freedom they provide to the writer to determine the extent and nature of their request (Reader 1991, p. 46). This flexibility is also based on the material characteristics and the anonymity of writing *ema*. The material characteristics (shape, size, placing of the printed illustrations) afford various kinds of ways to personalize *ema*—in anime pilgrimage, the emphasis is on the fans' drawings. At Seimei Shrine, the fans personalize the tablets by writing messages and decorating them with drawings and stickers in order to pray for Hanyu's success (Figure 2). In addition, anonymity can be seen as a factor in the flexibility of *ema*, as while some people write their full name, many only write their first name or initials, while some do not write their name at all. These factors enable fans to express themselves freely.

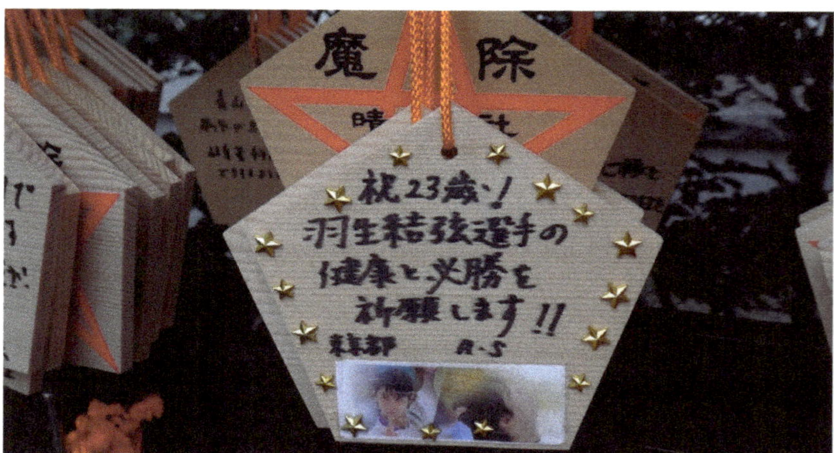

Figure 2. A decorated *ema* that wishes for the health and victory of the figure skater Yuzuru Hanyu.

Furthermore, through *ema* writing, the shrine becomes not only a shrine for Abe no Seimei, but through the fan's performances, it also enters a new network of sacred sites for Hanyu Yuzuru. In fact, the skater's presence can be seen in the votive tablets in other shrines dedicated to Abe no Seimei, such as the Abe no Seimei Shrine in Osaka.

Sometimes material objects themselves act as an attraction, as in the case of seal stamps (*goshuin*). Originally acquired as a proof of praying at a shrine or a temple, visitors collect the stamps in a special stamp book (*goshuin chō*) that one can purchase at temples or shrines.

Recently, the collection of seal stamps (*goshuin atsume*) has become a popular hobby and a reason to visit shrines and temples. In fact, many consider collecting *goshuin* a 'must' when visiting a shrine, as one can observe from the several guide books and websites dedicated to the topic. These kinds of media representations provide an example of how to perform at shrines (Okamoto 2020). In addition, due to the popularity of *goshuin* collection, one can purchase the stamp books in stationary shops and bookstores—this reflects how religious objects are produced and consumed outside of the original religious setting.

For many, the *goshuin* is the reason why they visit Seimei Shrine. One family, for example, looked up other tourist attractions around Nijō castle, and chose to visit Seimei Shrine because of the *goshuin*—they only had a vague image of the place before. Thus, while the shrine was not the primary attraction, *goshuin* collection acted as a pull factor for the visit. It can be said that the performance of *goshuin* collection produces a flow of material objects and people, as visitors move from one shrine to another. Another prominent feature of *goshuin* collection is photography. As the stamp as well as the stamp book in Seimei Shrine features the pentagram symbol, many want to take a picture at the shrine. Figure 3 shows how a visitor takes a photograph of the stamp book with the pentagram-shaped well in the background. This way, the visits are materialized both within the stamp books, which visitors use to remember their visits, and in photographs of the actual place.

Figure 3. Visitor taking a picture of the *goshuin* stampbook with the Seimei well.

However, the materiality of the stamp raises the question of authenticity. While some of the parts of the *goshuin* are stamped, usually the name of the shrine is written by hand. In Seimei Shrine, however, the name is also stamped, which replicates the characters written by Haretake Tsuchimikado, Abe no Seimei's descendant and a prominent figure in the history of Onmyōdō in the 1800s. Therefore, the characters are seen as meaningful in the context of the history of the shrine. While the meaning of the characters is explained on the pamphlet one receives when acquiring the *goshuin*, the fact that the characters are not written by hand has been criticized by some visitors. A comment on TripAdvisor says:

> "(Seimei Shrine is) a power spot I have wanted to visit. I was surprised by the constant flow of visitors. I wanted to get the *goshuin* but it was pre-written."
>
> (Posted by Eightwalker 2018)

For this visitor, the fact that the *goshuin* was not handwritten led them to the decision not to acquire one. On the other hand, the following comment shows the perceived importance of the appearance of the *goshuin*:

> "No part of the *goshuin* was handwritten, even the date was a stamp. At least they were honest about it when they confirmed whether it was ok that it was all stamped. You cannot tell whether it is handwritten or not from the quality of the *goshuin*. At least is better than the poor *goshuin* written by a part-timer with a dried up calligraphy pen at the famous Tenjin-san[14]."
>
> (Posted by (418ken 2017))

From these comments, one can see that the visitors have their own criteria of authenticity apart from the religious context. From the shrine's point of view, the *goshuin* are authentic—handwritten or not. However, the visitors think that the actual act of writing is a vital part of the authenticity of the *goshuin*. This can be said to reflect the way visitors perceive shrines and the objects there as something traditional, opposed to something seemingly premade or mass produced, such as a stamped *goshuin*. The fact that the shrine staff ask whether it is ok that the *goshuin* is stamped or not, shows that the shrine has also adapted to the visitors' needs.

Although the consciousness of belonging to one's local shrine has weakened, people tend to connect and find meaning in other shrines through their personal interests and memories. Some visitors talked about their personal connection to the shrine by sharing their life histories. These kinds of performances of personal remembrance are often expressed through material objects. A visitor from Tokyo, for example, talked about how they turned to Seimei Shrine when they had trouble at work:

> As I said before, I got a favor from the shrine a while ago. When I started working at my company, I had an unkind colleague. When I bought a talisman and placed it beneath my desk every day, the colleague went away [from the company]. I was very grateful because of that. Based on this experience, I can say that [the shrine] is very effective.
>
> (Visitor in their 40s)

Originally bought for a certain need, the talisman helped the visitor to overcome an obstacle, which then acts as evidence of the shrine's power. The visitor's action reflects the custom that religious objects, such as amulets and talismans, are usually brought back to the shrine where they have been purchased. In addition, if the visit or object has been beneficial, some people pay another visit to show their gratitude (*orei mairi*). For this visitor, however, visiting Seimei Shrine has also become a part of their routine in Kyoto, and the story of the talisman acts as a link between the visitor and the shrine. Although the visitor does not have the talisman anymore, the experience brings them back, 'contributing to their narratives and performances of self' (N. Morgan and Pritchard 2005, p. 45). The story of the talisman connects the visitor and the shrine, past and present events, as well as giving meaning to future visits.

These examples of the performances and views on votive tablets, seal stamps and talismans reflect how people connect with Seimei Shrine. While the shrine provides a certain framework, it also provides a setting for a range of other interpretations and performances, as visitors actively engage with it and its objects. It can be said that visitors understand the shrine through performances of popular and personal remembrance, which is materialized in the interactions between people and objects. These interactions reflect national trends (Porcu 2013), such as the *goshuin* collection and power spot boom. New meanings and interpretations are emplaced and enhanced through visitor performances

[14] It is not clear which shrine the commentor is referring to, as the name Tenjin-san is often used to refer to shrines of Sugawara no Michizane (845–903), nowadays revered as Tenman-Tenjin, the god for learning. From the context, the comment could be referring to Kitano Tenmangū Shrine in Kyoto, located fairly close to Seimei Shrine.

They also connect individual shrines to broader networks of people and goods. Sometimes these performances lead to conflicts, as I will discuss in the next section.

4.3. Contesting the Shrine—The Souvenir Dispute

As Chidester and Linenthal (1995) argued, sacred space is not inherently sacred. Instead, the sacredness of the space is constructed and therefore in constant flux. This can be seen in the way Seimei Shrine has actively adapted and participated in tourism instead of rejecting the outside media representations and new visitors. However, as new meanings are attached to the place through representation and contestation, the production of sacred space raises questions about authenticity and the process of authentication.

The discussion of authenticity has been central to tourism studies since the 1970s (MacCannell 1973; Cohen 1979) and has developed from object-based authenticity to a social constructivist approach (Bruner 2005), and further to discussions of existential authenticity (Wang 1999). However, lately, the focus has shifted from authenticity to the process of authentication. The study of authentication is the study of the social and political processes behind authenticity, i.e., the social process by which something (thing, site) is deemed real, true, or authentic (Cohen and Cohen 2012). As Cohen and Cohen (2012) discuss, authentication is not only based on formal authority and certification but also the informal performative processes of creating, preserving and reinforcing authenticity. Thus, not only official certifications, but also the practices of visitors can be seen as a means of authentication. In addition, as Jones (2010) in the context of material culture argues, authenticity is a product of the relationships between people, objects, and places—therefore, authentication is produced through the meanings people give, as well as embodied practices involving objects.

Souvenirs are an essential part of tourism. Seimei Shrine has also produced religious and touristic paraphernalia, which can be purchased at the shrine office and the official souvenir shop Kikyōan[15]. However, in 2011, Seimei Shrine appeared in the newspapers, as it had taken action against a neighboring souvenir shop called Onmyōji honpo. The unofficial souvenir shop had produced and sold souvenirs, such as good luck charms and a small Abe no Seimei mascot character called Seimei-kun. This led to Seimei Shrine publicly accusing the souvenir shop of the desecration of the deity revered in the shrine. The following announcement was posted on the shrine's official website on the 27th of October, 2011. Signs with the same contents were also placed within the shrine premises.

> The 'Onmyōji goods' and 'Good luck goods' sold close to the shrine by the Tajima orimono kabushikigaisha[16] (Onmyōji honpo), have not been purified or blessed at the shrine. We have nothing to do with them.
>
> Selling these kinds of objects close to the shrine is inappropriate and desecrates the divine virtues of Abe no Seimei considerably and we have asked the shop to stop selling them. However, the problem has not been solved and the situation is getting serious. For everyone visiting the shrine, please understand the situation and be careful of the following things.
>
> 1. Do not use the 'Onmyōji goods' or 'Good luck goods' when visiting the shrine.

15 Miller has noted that Shinto shrines dedicated to Abe no Seimei started producing and selling religious paraphernalia targeted especially towards young girls during the Seimei boom (Miller 2008, p. 36). Seimei Shrine sells both souvenirs and religious objects within its premises. The religious objects are mainly sold at the shrine office, whereas souvenirs are sold at Kikyōan. At the time of my fieldwork, religious objects, such as 16 different types of amulets (*omamori*), three types of talismans (*ofuda*), and purifying sand were sold at the shrine office. There are also special amulets, only available during certain periods or events. However, the shrine office also sells seemingly secular items such as a small comic book about Abe no Seimei's life, as well as special stationery.

16 The textile company, which owned the souvenir shop.

2. These 'goods' cannot be called *ofuda, omamori* (amulets, charms). Do not place them in the return box[17].
3. We might inquire about these acts from a legal and religious position. We ask for your cooperation.

(Seimei Shrine 2012)

From the official announcement, we can see that souvenirs and their authenticity are in the center of this conflict. The official announcement also tells us that there are three different actors involved in the dispute—the shrine, the souvenir shop and the visitors. All of these actors involved had their own meanings and sense of authenticity, which is constituted in different ways.

First of all, the statement used religious rhetoric to emphasize the shrine's authority over the objects as well as Abe no Seimei. The unofficial nature of the souvenirs was based on the fact that they had not been officially purified, blessed or produced by the shrine and therefore the souvenirs were seen as desecrating and offending Abe no Seimei. The souvenirs were also seen as dangerous, as the description of the Seimei Shrine's official souvenir shop seems to imply: 'We sell memorabilia and souvenirs at Kikyōan. As everything is authorized by Seimei Shrine, one can purchase them *with peace of mind*.' (Italics added by the author.)

The authenticity of the objects was also related to the practices they were involved in. The souvenir shop, for example, sold the souvenirs by using Abe no Seimei's story in order to make profit. According to the owner, they 'just wanted to spread Abe no Seimei to the world by making him into a cute character' (Murakami 2012). From their perspective, the souvenirs were not even meant for religious practices. While many shrines and temples are using mascot characters in their branding, (see Porcu 2014), the shrine decided not to embrace the new character. From the shrine's point of view, Seimei-kun and the other souvenirs had a blasphemous and non-authentic status. This was enhanced further by the fact that the visitors had used the unofficial souvenirs as religious objects within the shrine premises, as implied in the official announcement. From the visitors' point of view, though, the souvenirs were a part of the overall experience at the shrine. A souvenir can become a religious object and vice versa, as the same object can have multiple meanings and uses during its lifetime (Kopytoff 1986; N. Morgan and Pritchard 2005).

Through their performances involving the unofficial souvenirs, the visitors unconsciously contested the shrine's official meanings and interpretations, as well as its authority. The shrine reacted to this by controlling and banning unwanted behavior by using its authoritative status within the religious context. However, by mentioning the possible legal consequences, the announcement implies that the souvenirs were seen as a problem from not just a religious perspective but also a secular one. In fact, the official souvenir shop called Kikyōan sells souvenirs similar to the ones of Onmyōji honpo. However, the official souvenirs have a sticker of approval to distinguish their authenticity. From this, it is clear that the unofficial souvenir shop was a business rival, which threatened Seimei Shrine's brand. That is why the problem was not the inherent authenticity of the objects but the authority over who is able to tell the story of the shrine (Bruner 2005), whether it be the shrine, the souvenir shop, or the visitors. In the end, Onmyōji honpo closed its store in 2016, but still continues to sell its items online.

The dispute can be seen as an example of how the dynamics of the shrine changed through the involvement of tourism. When the shrine and souvenirs enter the touristic context, the authority of the shrine as well as the status of the objects become relative. While the physical movement of the objects as well as the involvement in the visitors' practices were the igniting factors, within the dispute the objects became a part of the authentication process—a way to show authority and control the interpretations and

17 The *osamefudadokoro* is a box in which old religious objects are placed when brought back to the shrine. There is a custom that old religious objects should be ritually disposed of at a shrine. Many of the visitors interviewed in January 2018 brought back their old amulets to Seimei Shrine.

practices of visitors. However, the mixing and matching of different narratives and practices shows that within the individuals' experiences they themselves are the upmost authority regarding authenticity, which reflects the individualization of modern religion as well as tourism.

5. Discussion

The relationship between religion and tourism reflects the change and transformation of religion in the late-modern consumer society, which can be seen in both the way in which religious institutions adapt to, and visitors understand, religion. By focusing on material objects and the interactions they are involved in, one can observe how the boundaries between the religious and the secular are shifting in individual performances as well as the shrine itself.

One way to explain this is through the metaphor of a spiritual market, proposed by Yamanaka (Yamanaka 2016, 2017, 2020). According to Yamanaka, in the traditional religious market, religions used to offer goals which are considered demanding to attain, such as salvation and pilgrimage. Opposed to this, nowadays, consuming religion is linked to so called 'self-reflection' or 'finding oneself'—themes which are more easily approachable and relatable to modern people. The needs which this consumption establishes are then fulfilled with different products, such as those of religious tourism. Referring to Zygmunt Bauman's notions of liquid modernity and light capitalism, Yamanaka calls this phenomenon light religion (*karui shūkyō*). The lightness refers to a situation, where the religious and non-religious markets merge, and religious ideas and practices are displaced and consumed outside of their original contexts within a new spiritual market. In the case of Seimei Shrine, this religious transformation can be seen in the way the shrine has changed its image through branding and commodification, as well as accommodating the visitors' various needs. These needs are reflected in the material environment of the shrine as well as the visitor performances, in which popular culture and media representations have permeated.

While most interviewees mentioned Seimei Shrine's connection to popular culture, some visitors came for the shrine's religious benefits or services. As Seimei Shrine is famous for preventing danger (*yakuyoke*), some visitors chose the shrine because it was their dangerous year (*yakudoshi*)[18]. However, the visitors interviewed did not participate in any special rituals and their visit did not differ from the average tourist's. Seimei Shrine also offers life counseling services (*jinsei sōdan*), which are based on fortunetelling. This is peculiar to Seimei Shrine, as not all Shinto shrines offer this kind of service. Some visitors had come to the shrine because of that (some even specified that they had found the shrine when searching for '*uranai*', which means fortunetelling).

For most, however, the visits were not part of a 'finding oneself' narrative—instead, most of the visitors enjoyed the peculiarity of the shrine's material environment as well as the narratives attached to it. In fact, most of the visitors did not see the visit to Seimei Shrine as religious nor spiritual, and it was understood in a touristic context. From an etic point of view, they however participated in seemingly religious practices, such as prayers and *goshuin* collection. One reason for this, as Nakanishi (2018) argues, is that Shinto has permeated in the daily lives of the Japanese and visiting a shrine is often not consciously recognized as a form of religion[19].

What makes visitors engage in these kinds of practices at Seimei Shrine? While some researchers have argued that practices reflect so-called 'affective belief', which functions on an emotional level, as opposed to cognitive or doctrine-based belief (Reader and Tanabe

18 *Yakudoshi* are certain ages when people are thought to be especially open to misfortune. The years before and after are also considered dangerous. According to Reader and Tanabe, 'praying for the eradication of dangers is a major category of benefit seeking, and *yakudoshi*-related prayers and actions are one of the most prevalent occasions when this occurs' (Reader and Tanabe 1998, p. 62).

19 From a historical point of view, this can also be linked to the separation of the religious and the secular in the modernization of Japan, especially with the formation and use of the term religion (*shūkyō*). As a political choice, to distinguish Shinto (or specifically State Shinto) from Buddhism and Christianity, Shinto was classified as morality instead of a religion (Isomae 2012). As an in-depth discussion of the topic is beyond the scope of this article, see for example (Isomae 2012, 2014; Josephson 2012; Krämer 2013; Rots and Teeuwen 2017).

1998), in the study of material religion, the assumption of underlying belief has been problematized. Keane (2008) argued against taking practices or material objects as expressions or evidence of prior beliefs. Okamoto (2019b) also talked about how seemingly religious practices at power spots cannot be taken as evidence for a new religious consciousness, as there is no set belief system behind the practices. Instead, the main reason behind the practices seems to be whether or not the places enable these kind of practices (Okamoto 2019b, p. 98). Additionally, Kadota (2016) discussed how the instructions and management techniques at sacred sites produce so-called artificial belief (*jinkōteki shinkō*), which can be seen in the form of prayers in places where one is *supposed* to pay one's respects. Additionally, in Seimei Shrine, it can be said that most just follow the examples of the different media representations and instructions provided at the shrine. Many touched the sacred tree because they were *supposed* to—in the context of power spots.

Okamoto and Kadota discuss the practices as reactions to the material environment. As we have seen in the examples in Part 4, objects such as the seal stamps and *ema* act as intermediaries through which the visitors connect with Abe no Seimei, the shrine, as well as other people, including their idols. However, while the meanings are varied, the performances themselves are not something completely new. To explore how mediations bind and bond believers with each other and to the transcendental, Birgit Meyer has coined the notion of 'sensational form', by which they mean 'relatively fixed, authorized modes of invoking and organizing access to the transcendental, thereby creating and sustaining links between believers in the context of particular religious power structures' (Meyer 2009, p. 13). These include both religious content (such as beliefs, doctrines, sets of symbols) and norms. According to Meyer, 'sensational forms can best be understood as a condensation of practices, attitudes, and ideas that structure religious experiences and hence "ask" to be approached in a particular matter' (Meyer 2009, p. 13). As mentioned above, while Shinto shrines do not have certain doctrines, there are certain ways to interact with the material objects within the shrine.

While Meyer's discussion is focused on new media, tourism can also be seen as a phenomenon which has an impact on established sensational forms. In fact, many people engage with and understand religious themes and narratives through tourism and popular culture, which is apparent in the case of Seimei Shrine. While Porcu has noted that 'the market and media have acquired a privileged position in creating meaning for contemporary religious practices' (Porcu 2013, p. 292), tourism is involved in the reconstruction and negotiation of practices, not only meanings. As we have seen, tourism brings about images and expectations which are projected onto the material objects and the performances involving them. Within these interactions, as Rambelli has noted, 'sacred objects function as interfaces, and the realm of materiality in general can be defined as a space of interplay between the secular, the sacred, and their respective economies (systems of production, exchange, and representation)' (Rambelli 2007, p. 273). Value is also important, such as the multiple criteria of authenticity seen in the examples of stamped *goshuin* and the souvenir dispute.

While it is true that 'the relevance of the shrines is not in their religious messages but in the considerable freedom they provide individuals to use its precincts for different pursuits' (Nelson 1996, p. 117), in the case of Seimei Shrine, the relevance seems to lie in the active adaptation to new meanings and situations. Not just passively providing a set for visitor performances, the shrine has actively changed its material environment and image to suit the needs of the visitors. It also encourages the visitors to connect physically with the objects within the shrine. Through these interactions, the shrine and its objects provide the visitors with feelings of enjoyment and sometimes also connectedness, whether it is to a deity, a favorite character in a movie, or a community of fans. These experiences are enhanced by different mediations, be it the media representations or tangible amulets bought at the shrine. For fans of Yuzuru Hanyu, visiting Seimei Shrine and purchasing religious items are a way to express one's identity as a fan, not necessarily expressing an underlying belief.

Furthermore, tourism also brings about contestation and conflict, as seen in the case of the souvenir dispute. In Chidester and Linenthal's view, religious actors are often opposed to the entanglement of so-called 'profane' forces (Chidester and Linenthal 1995, p. 17), which was also the case with Jinja Honchō's response to the power spot boom. However, as we can see in the case of Seimei Shrine, not all religious actors are necessarily opposed to these forces, as many religious institutions are responding to national and global trends by actively participating in various aspects of tourism. Seimei Shrine has unarguably done this and much more—it has established itself as a popular tourist attraction in Kyoto.

Yet, as the boundaries of religious and secular institutions are becoming fluid through the involvement of tourism, religious institutions have to face situations which they have not had to worry about before. Different actors are bound to get involved in religious places, and as Kinnard argues, 'it is precisely the competing conceptions and motivations of the various actors involved in a particular place . . . that constitute it' (Kinnard 2014, p. 2). While the shrine has embraced a seemingly 'soft' image through its connections to popular culture (Porcu 2014), during the souvenir dispute it pleaded to its status as a religious actor to protect its status. This goes against the fact that the shrine does not normally impose a strict doctrine on its visitors. This shows that the shrine has entered a new touristic market, in which it has to protect its status in order to differentiate itself from other shrines as well as other tourist attractions of Kyoto. In this context, the religious aspect is just one of the many meanings of the shrine—Seimei Shrine has to balance its attractiveness as a tourist attraction as well as its integrity as a religious institution.

6. Conclusions

In this paper, I have analyzed tourism at Seimei Shrine through the three aspects of constructing, performing, and contesting sacred space. Since the 2000s, Seimei Shrine has adapted to the touristic context through the theming of its premises as well as its media presence. With the use of material objects, such as bronze statues and other paraphernalia, the shrine has adapted to the outside media representations and the visitors' expectations. This can be seen as a reflexive process, where the shrine has had to re-evaluate its position within the religious as well as the touristic market. By embracing tourism, the shrine has adapted to the late-modern consumer society, in which it is no longer the center for religious life but has to compete with other shrines and tourist attractions in Kyoto.

Visitors consume the shrine through different performances. It became clear that while there is not much variety in the practices, the interpretations and meanings are connected to personal and popular remembrance. While some understand the material objects through personal memories, for many, the context in which the shrine is consumed is highly mediatized. Thus, the material objects provide an interface for self-expression and the construction of personal ties, as different fandoms have found their place within the shrine. Therefore, the case of Seimei Shrine reflects the overlap between religion and popular culture in a concrete way.

It also became clear that Seimei Shrine has several roles—as a religious actor, as well as a tourist attraction. The souvenir dispute revealed that Seimei Shrine is not only a vessel for different outside representations; it is also an active participant in the process. While the situation brings about new challenges regarding authenticity as well as authority, the involvement in tourism can be seen as a way to respond to the shift of the place of religion in modern society as well as the diverse needs of the visitors. Another great challenge for Shinto shrines is the COVID-19 pandemic, as festivals are cancelled (Amada 2020) and visitors are not physically able to visit the shrine. Moreover, many shrines, including Seimei Shrine, have changed the shape of the *temizusha* so that the visitors do not have to use a ladle when purifying themselves. While seemingly small, this change affects the way visitors interact with the material environment and the objects within the shrine. That is why the exploration of the shrine's secular connections as well as the material changes they impose can be seen as the key to placing Shinto shrines in the context of wider societal change.

Funding: The fieldwork which this paper is based on, was conducted when the author was a recipient of the Hokkaido University President's Scholarship in 2017-2019. The APC was partly funded by the Hokkaido University Graduate School of International Media, Communication, and Tourism Studies.

Informed Consent Statement: Informed consent was obtained from all subjects involved in the study.

Data Availability Statement: The data presented in this study are available on request from the corresponding author.

Conflicts of Interest: The author declares no conflict of interest. The funders had no role in the design of the study; in the collection, analyses, or interpretation of data; in the writing of the manuscript, or in the decision to publish the results.

References

418ken. 2017. "Goshuin Ga ... " TripAdvisor. Available online: https://www.tripadvisor.jp/ShowUserReviews-g298564-d1568228-r463974741-Seimei_Shrine-Kyoto_Kyoto_Prefecture_Kinki.html (accessed on 17 December 2020).

Amada, Akinori. 2020. Sairei No Chūshi, Yōkai No Ryūkō—'Ekibyō Yoke' Wo Tegakari Ni. In *Ajia Yūgaku 253: Posuto Korona Jidai No Higashi Ajia*. Edited by Mooam Hyun and Yohei Fujino. Tokyo: Bensei Shuppan, pp. 191–202.

Andrews, Dale K. 2014. Genesis at the Shrine: The Votive Art of an Anime Pilgrimage. In *Mechademia 9*. Minneapolis: University of Minnesota Press, pp. 217–33.

Asahi Shimbun. 2010. (Shun-Kan) Pawā, Kitemasu (Osaka). *Asahi Shimbun*, November 6.

Bærenholdt, Jørgen Ole, Michael Haldrup, Jonas Larsen, and John Urry. 2004. *Performing Tourist Places*. London: Ashgate.

Bruner, Edward M. 2005. *Culture on Tour: Ethnographies of Travel*. Chicago: The University of Chicago Press.

Bryman, Alan. 2004. *The Disneyization of Society*. London: SAGE Publications.

Carter, Caleb. 2018. Power Spots and the Charged Landscape of Shinto. *Japanese Journal of Religious Studies* 45: 145–74. [CrossRef]

Chidester, David, and Edward T. Linenthal. 1995. Introduction. In *American Sacred Space*. Edited by David Chidester and Edward T. Linenthal. Bloomington: Indiana University Press, pp. 1–42.

Cohen, Erik. 1979. A Phenomenology of Tourist Experiences. *Sociology* 13: 179–201. [CrossRef]

Cohen, Erik, and Scott A. Cohen. 2012. Authentication: Hot and Cool. *Annals of Tourism Research* 39: 1295–314. [CrossRef]

Collins-Kreiner, Noga. 2020. A Review of Research into Religion and Tourism Launching the Annals of Tourism Research Curated Collection on Religion and Tourism. *Annals of Tourism Research* 82: 102892. [CrossRef]

Daniels, Inge. 2003. Scooping, Raking, Beckoning Luck: Luck, Agency and the Interdependence of People and Things in Japan. *Journal of the Royal Anthropological Institute* 9: 619–38. [CrossRef]

Daniels, Inge. 2012. Beneficial Bonds: Luck and the Lived Experience of Relatedness in Contemporary Japan. *Social Analysis* 56: 148–64. [CrossRef]

Edensor, Tim. 1998. *Tourists at the Taj. Performance and Meaning at a Symbolic Site*. London: Routledge.

Edensor, Tim. 2000. Staging Tourism: Tourists as Performers. *Annals of Tourism Research* 27: 322–44. [CrossRef]

Edensor, Tim. 2001. Performing tourism, staging tourism: (Re)producing tourist space and practice. *Tourist Studies* 1: 59–81. [CrossRef]

Eightwalker. 2018. "Pawāsupotto." TripAdvisor. Available online: https://www.tripadvisor.jp/ShowUserReviews-g298564-d1568228-r639327485-Seimei_Shrine-Kyoto_Kyoto_Prefecture_Kinki.html (accessed on 17 December 2020).

Einstein, Mara. 2007. *Brands of Faith: Marketing Religion in a Commercial Age*. New York: Routledge.

Foster, Michael Dylan. 2020. Eloquent Plasticity. *Journal of Religion in Japan* 9: 118–64. [CrossRef]

Haldrup, Michael, and Jonas Larsen. 2010. *Tourism, Performance and the Everyday: Consuming the Orient*. London: Routledge.

Hayashi, Makoto, and Matthias Hayek. 2013. Editors' Introduction: Onmyodo in Japanese History. *Japanese Journal of Religious Studies* 6: 1–6. [CrossRef]

Higgins, Leighanne, and Kathy Hamilton. 2020. Pilgrimage, Material Objects and Spontaneous Communitas. *Annals of Tourism Research* 81: 102855. [CrossRef]

Hirayama, Noboru. 2015. *Hatsumōde No Shakai Shi. Tetsudō Ga Unda Goraku to Nashonarizumu*. Tokyo: Tokyo Daigaku Shuppan-kai.

Hirschkind, Charles. 2006. *The Ethical Soundscape: Cassette Sermons and Islamic Counterpublics*. New York: Columbia University Press.

Horie, Norichika. 2017. The Making of Power Spots: From New Age Spirituality to Shinto Spirituality. In *Eastspirit: Transnational Spirituality and Religious Circulation in East and West*. Leiden: BRILL, pp. 192–217. [CrossRef]

Horie, Norichika. 2019. *Poppu Supirichuariti: Media Ka Sareta Shūkyō*. Tokyo: Iwanami Shoten.

Industry and Tourism Bureau. 2019. "Kyōto Kankō Sōgō Chōsa." City of Kyoto. Available online: https://www.city.kyoto.lg.jp/sankan/page/0000271459.html (accessed on 23 December 2020).

Ishii, Kenji. 2010. Purorōgu: Jinja Shintō Wa Suitai Shita No Ka. In *Shintō Wa Doko E Iku Ka*. Edited by Kenji Ishii. Tokyo: Perikansha, pp. 11–30.

Isomae, Jun'ichi. 2012. The Conceptual Formation of the Category 'Religion' in Modern Japan: Religion, State, Shintō. *Journal of Religion in Japan* 1: 226–45. [CrossRef]

Isomae, Jun'ichi. 2014. Religious Discourse in Modern Japan: Religion, State, and Shintō. Leiden: BRILL.

Jones, Siân. 2010. Negotiating Authentic Objects and Authentic Selves. *Journal of Material Culture* 15: 181–203. [CrossRef]

Josephson, Jason Ānanda. 2012. *The Invention of Religion in Japan*. Chicago: The University of Chicago Press.
Kadota, Takehisa. 2013. Junrei Tsūrizumu No Minzoku Shi—Shōhi Sareru Shūkyō Keiken. Tokyo: Shinwasha.
Kadota, Takehisa. 2016. Seichikankō No Kūkanteki Kōchiku–Okinawa, Sefa-Utaki No Kanrigihō To "Seichirashisa" No Seisei Wo Megutte. *Kankogakuhyōron* 4: 161–75.
Keane, Webb. 2007. *Christian Moderns: Freedom & Fetish in the Mission Encounter*. Berkeley: University of California Press.
Keane, Webb. 2008. On the Materiality of Religion. *Material Religion* 4: 230–31. [CrossRef]
King, E. Frances. 2010. *Material Religion and Popular Culture*. New York: Routledge.
Kinnard, Jacob N. 2014. *Places in Motion: The Fluid Identities of Temples, Images, and Pilgrims*. New York: Oxford University Press.
Kopytoff, Igor. 1986. The Cultural Biography of Things: Commoditization as Process. In *The Social Life of Things*. Cambridge: Cambridge University Press, pp. 64–92.
Krämer, Hans Martin. 2013. How 'Religion' Came to Be Translated as Shukyo: Shimaji Mokurai and the Appropriation of Religion in Early Meiji Japan. *Japan Review* 25: 89–111.
Larsen, Jonas. 2006. Geographies of Tourism Photography: Choreographies and Performances. In *Geographies of Communication: The Spatial Turn in Media Studies*. Edited by André Jansson and Jesper Falkheimer. Gothenburg: Nordicom, pp. 243–60.
MacCannell, Dean. 1973. Staged Authenticity: Arrangements of Social Space in Tourist Settings. *American Journal of Sociology* 79: 589–603. [CrossRef]
MacCannell, Dean. 1999. *The Tourist. A New Theory of the Leisure Class*. Berkeley: University of California Press.
Meyer, Birgit, David Morgan, Crispin Paine, and S. Brent Plate. 2010. The Origin and Mission of Material Religion. *Religion* 40: 207–11. [CrossRef]
Meyer, Birgit. 2009. Introduction: From Imagined Communities to Aesthetic Formations: Religious Mediations, Sensational Forms, and Styles of Binding. In *Aesthetic Formations*. Edited by Birgit Meyer. New York: Palgrave Macmillan. [CrossRef]
Miller, Laura. 2008. Extreme Makeover for a Heian-Era Wizard. In *Mechademia 3*. Minneapolis: University of Minnesota Press, pp. 30–45.
Morgan, David, ed. 2010. *Religion and Material Culture: The Matter of Belief*. New York: Routledge.
Morgan, Nigel, and Annette Pritchard. 2005. On Souvenirs and Metonymy: Narratives of Memory, Metaphor and Materiality. *Tourist Studies* 5: 29–53. [CrossRef]
Motion Picture Producers Association of Japan Inc. 2001. 2001-Nen (Heisei 13-Nen) Kōshū 10-Oku En Ijō Bangumi. Available online: http://www.eiren.org/toukei/img/eiren_kosyu/data_2001.pdf (accessed on 23 December 2020).
Murakami, Koichi. 2012. Kyoto, Seimei Jinja vs. Seimei Kun. *Asahi Shimbun*, December 26.
Nakanishi, Yuji. 2018. Shintoism and Travel in Japan. In *Tourism and Religion: Issues and Implications*. Edited by Richard Butler and Wantanee Suntikul. Bristol: Channel View Publications, pp. 68–82.
Nelson, John K. 1996. Freedom of Expression: The Very Modern Practice of Visiting a Shinto Shrine. *Japanese Journal of Religious Studies* 23: 117–53. [CrossRef]
Nelson, John K. 2000. Freedom of Expression: The Very Modern Practice of Visiting a Shinto Shrine. In *Enduring Identities: The Guise of Shinto in Contemporary Japan*. Honolulu: University of Hawai'i Press, pp. 22–52.
Okamoto, Ryosuke. 2015. *Seichi Junrei—Sekai Isan Kara Anime No Butai Made*. Tokyo: Chuōkōron Shinsha.
Okamoto, Ryosuke. 2019a. *Pilgrimages in the Secular Age: From El Camino to Anime*. Tokyo: Japan Publishing Industry Foundation for Culture (JPIC).
Okamoto, Ryosuke. 2019b. Practicing without Believing in Post-Secular Society: The Case of Power Spot Boom in Contemporary Japan. *Jōchi Daigaku Yōroppa Kenkyū Sōsho* 12: 94–99.
Okamoto, Ryosuke. 2020. Jenerikku Shūkyō Shiron. In *Gendai Shūkyō To Supirichuaru Māketto*. Edited by Hiroshi Yamanaka. Tokyo: Kōbundō, pp. 27–46.
Paine, Crispin. 2019. Beyond Museums: Religion in Other Visitor Attractions. *ICOFOM Study Series* 47: 157–69. [CrossRef]
Porcu, Elisabetta. 2012. Observations on the Blurring of the Religious and the Secular in a Japanese Urban Setting. *Journal of Religion in Japan* 1: 83–106. [CrossRef]
Porcu, Elisabetta. 2013. Sacred Spaces Reloaded: New Trends in Shinto. In *Self-Reflexive Area Studies*. Edited by Matthias Middell. Leipzig: Leipziger Universitätsverlag, pp. 279–94.
Porcu, Elisabetta. 2014. Pop Religion in Japan: Buddhist Temples, Icons, and Branding. *Journal of Religion and Popular Culture* 26: 157–72. [CrossRef]
Rambelli, Fabio. 2007. *Buddhist Materiality: A Cultural History of Objects in Japanese Buddhism*. Stanford: Stanford University Press.
Reader, Ian, and George Tanabe. 1998. *Practically Religious: Worldly Benefits and the Common Religion of Japan*. Honolulu: University of Hawai'i Press.
Reader, Ian. 1991. Letters to the Gods: The Form and Meaning of Ema. *Japanese Journal of Religious Studies* 1: 24–50. [CrossRef]
Reader, Ian. 2014. *Pilgrimage in the Marketplace*. London: Routledge. [CrossRef]
Rots, Aike P. 2017. Public Shrine Forests? Shinto, Immanence, and Discursive Secularization. *Japan Review* 30: 117–53.
Rots, Aike P., and Mark Teeuwen. 2017. Introduction: Formations of the Secular in Japan. *Japan Review* 30: 3–20.
Sano, Kazufumi. 2007. Ujiko. In *Encyclopedia of Shinto*. Tokyo: Kokugakuin University, Available online: http://eos.kokugakuin.ac.jp/modules/xwords/entry.php?entryID=1161 (accessed on 23 December 2020).

Seimei Shrine. 2012. "Seimei Jinja Yori Taisetsuna Oshirase [Important Announcement from Seimei Shrine]." 2012. Available online: https://www.seimeijinja.jp/archives/date/2011/10?cat=3 (accessed on 8 October 2018).

Shigeta, Shin'ichi. 2013. A Portrait of Abe No Seimei. *Japanese Journal of Religious Studies* 40: 77–97. [CrossRef]

Shinde, Kiran A. 2013. "Re-Scripting the Legends of Tuḷjā Bhavānī: Texts, Performances, and New Media in Maharashtra". *International Journal of Hindu Studies* 17: 313–37. [CrossRef]

Shinde, Kiran A. 2020. Religious Theme Parks as Tourist Attraction Systems. *Journal of Heritage Tourism*, 1–19. [CrossRef]

Sonoda, Minoru. 1975. The Traditional Festival in Urban Society. *Japanese Journal of Religious Studies* 1: 103–36. [CrossRef]

Suga, Naoko. 2010. Pawāsupotto to Shite No Jinja. In *Shinto Wa Doko e Iku Ka*. Edited by Kenji Ishii. Tokyo: Perikansha, pp. 232–52.

Suzuki, Ryotaro. 2020. Shōhi Sareru Basho, Datsu Shōhin Ka Sareru Tabi: Pawāsupotto Kankō to Ryokō Sangyō No Henyō. In *Gendai Shūkyō To Supirichuaru Māketto*. Edited by Hiroshi Yamanaka. Tokyo: Kōbundō, pp. 47–66.

Takaba, Mizuho. 2015. Hanyu Narikitta Onmyoji Sekai Ni 'wa' Mitomesaseta. *Nikkan Sports*. November. Available online: https://www.nikkansports.com/sports/news/1572765.html (accessed on 23 December 2020).

Terzidou, Matina. 2020. Re-Materialising the Religious Tourism Experience: A Post-Human Perspective. *Annals of Tourism Research* 83: 102924. [CrossRef]

Timm Knudsen, Britta, and Anne Marit Waade. 2010. Performative Authenticity in Tourism and Spatial Experience: Rethinking the Relations between Travel, Place and Emotion. In *Re-Investing Authenticity: Tourism, Place and Emotions*. Edited by Anne Marit Waade and Britta Timm Knudsen. Bristol and Buffalo: Channel View Publications, pp. 1–20.

Tsukada, Hotaka, and Toshihiro Omi. 2011. Gendai Nihon 'Shūkyō' Jōhō No Hanran. Shin Shūkyō, Pawāsupotto, Sōgi, Butsuzō Ni Chūmoku Shite. In *Gendai Shūkyō 11: Gendai Shūkyō No Naka No Shūkyō Dentō*. Tokyo: Akiyama Shoten, pp. 284–307.

Urry, John, and Jonas Larsen. 2011. *The Tourist Gaze 3.0*. London: SAGE Publishing.

Urry, John. 1990. *The Tourist Gaze: Leisure and Travel in Contemporary Societies*. London: SAGE Publications.

Vásquez, Manuel A. 2011. *More Than Belief: A Materialist Theory of Religion*. New York: Oxford University Press.

Wang, Ning. 1999. Rethinking Authenticity in Tourism Experience. *Annals of Tourism Research* 26: 349–70. [CrossRef]

Yamamura, Takayoshi. 2015. Contents Tourism and Local Community Response: Lucky Star and Collaborative Anime-Induced Tourism in Washimiya. *Japan Forum* 27: 59–81. [CrossRef]

Yamanaka, Hiroshi. 2012. Shūkyo To Tsūrizumu Kenkyū Wo Mukete. In *Shūkyō To Tsūrizumu*. Edited by Hiroshi Yamanaka. Kyoto: Sekaishishosha, pp. 3–30.

Yamanaka, Hiroshi. 2015. Tsūrizumu To Konnichi No Seichi—Nagasaki No Kyōkaigun No Sekai Isan Ka Wo Chūshin Ni Shite. *Shigaku* 85: 581–610.

Yamanaka, Hiroshi. 2016. Shūkyō Tsūrizumu To Gendai Shūkyō. *Kankogakuhyōron* 4: 149–59.

Yamanaka, Hiroshi. 2017. Shōhishakai Ni Okeru Gendai Shūkyō No Henyō. *Shūkyō Kenkyū* 90: 255–80.

Yamanaka, Hiroshi. 2020. Gendai Shūkyō To Supirichuaru Māketto. In *Gendai Shūkyō To Supirichuaru Māketto*. Edited by Hiroshi Yamanaka. Tokyo: Kōbundō, pp. 1–26.

Article

World Youth Day 2016 in the Archdiocese of Lodz: An Example of the Eventization of Faith

Joanna Bik and Andrzej Stasiak *

University of Lodz, Faculty of Geographical Sciences, Institute of Urban Geography and Tourism Studies, 31 Kopcińskiego Str, 90-142 Łódź, Poland; joannabik93@gmail.com

* Correspondence: andrzej.stasiak@geo.uni.lodz.pl

Received: 25 June 2020; Accepted: 27 September 2020; Published: 1 October 2020

Abstract: The organization of numerous religious mass events of international, or even global, reach is a phenomenon of the early 21st century. It is sometimes termed "eventization of faith". This article presents a multifaceted analysis of the initial stage of the World Youth Day in 2016, which took place in the Archdiocese of Łódź (Poland). While multiple scholarly publications have been written about World Youth Day (WYD) itself, its first part of preparatory nature, the so-called "Days in Dioceses", has not been studied yet. The authors of this paper used a wide array of research methods, such as participant observation, questionnaire (official statistics concerning 10,000 pilgrims), pilot survey (258 respondents), and analysis of media reports (over 100 films and 30 articles). The analysis of the organizational method of such a major religious event leads to a conclusion that it is a complex logistic undertaking, which requires professional preparation and implementation by a team of specialists in different fields as well as an army of deeply involved volunteers and public services employees. Over 10.2 thousand young pilgrims (mostly at the age of 15–29) participated in the youth meeting in the Archdiocese of Łódź; apart from spiritual motives (strengthening faith, meeting Pope Francis, following in the footsteps of St. John Paul II) they exhibited strong social (willingness to be in the community of believers, making new friends), recreational and tourist (visiting Poland) needs as well. In view of the hermetic and low-budget character of World Youth Day, its impact on the economy of the region was deemed negligible. Above all, the event played a promotional and image-building part, which perhaps in the years to come will result in an increase in visits of foreign tourists to Łódź.

Keywords: World Youth Day; eventization of faith; religious tourism; Archdiocese of Lodz; Poland

---

1. Introduction

World Youth Day is a mass religious event open to the followers of other religions, non-believers or those searching for faith. The only condition for participation is to show respect for the Catholic character of the event and for the dignity of all people, regardless of their sex, origin, color or creed. The main aim of the event is to meet God, the Pope, other human beings, and oneself, and in this way build a community of Catholics—a community of young people guided by the same values (Bik 2016). From the organizers' point of view, World Youth Day (WYD) has one additional aim—to create a positive image of the host country. Therefore, the program of the pilgrims' stay includes tourist, cultural, recreational and other elements.

World Youth Day (WYD) is a meeting of young Catholics, an event initiated by Pope John Paul II in 1985 and held every 2–3 years. In other years, it is celebrated locally in all the dioceses of the

Catholic Church. So far, the international meetings have been organized 14 times[1], and including local events, over 30 times.

An exceptional feature of World Youth Day is the number of people taking part in it: from 300,000 to over 4 million. The biggest meeting ever took place in Manila in 1995 and was officially included in the Guinness Book of Records as the greatest gathering at a single location in the world. The 31st event was held on 26–31 August 2016 in Krakow (for the second time in Poland) and attracted ca. 2.5 million pilgrims.

WYD celebrations usually include 14 kinds of religious event which are held on a specific day or even one part of a day (e.g., welcoming the Pope, the Way of the Cross, all-night vigil, the final mass, volunteers' meeting with the Pope). In 1997, in Paris, the tradition of the "prologue" was started, called Days in the Dioceses. It is the week directly preceding World Youth Day which young pilgrims spend in the different dioceses of the host country. At this time, they get ready spiritually for the main religious celebrations, but also have time for sightseeing, learning about culture and traditions, as well as meeting the inhabitants of the host country (e.g., by staying in local families' homes). In 2016, the Days in the Dioceses were held on 19–25 July (in some years, in exceptional situations they started on 15 July), and one of the most visited places was the Archdiocese of Lodz (over 10,000 visitors).

2. Literature Review

"Eventization" is a relatively new term. The first mentions of a broader concept of event tourism appeared in the 1970s (Ritchie and Donald 1974). However, it was not until the 1980s and 1990s that this form of tourism had become a popular object of study conducted in particular by researchers from Australia, Canada and the USA (Gartner and Holecek 1983; Ritchie 1984; Hall 1989). Initially, event tourism was analyzed from an economic perspective. With time, researchers also became interested in the participants and the impact on spatial development while the 1990s brought a number of publications focusing primarily on management and its increasing professionalization (Getz 1991; Uysal et al. 1993). In the 21st century, there were studies on the aware and purposeful policy (strategy) of events (e.g., Foley et al. 2012), serving, among others, the regeneration and revitalization of cities (e.g., Richards and Robert 2010; Cudny 2016). Popular examples were various types of cultural and sports events such as the Olympic Games. Although Getz (2008) mentions religious events in his typology, he does not elaborate further on this topic.

"Faith eventization", as the contemporary organization of religious events is referred to, is an expression which appeared in the literature in the 21st century (Pfadenhauer 2010; Zduniak 2010, 2015; Bilska-Wodecka and Izabela 2014). The organization of spectacular events is a tool for implementing a well-thought-out marketing strategy, assuming improvement of the general image of the Church, promotion of religion and pilgrimage, and, consequently, gaining new followers (evangelization). However, since contemporary religiosity and pilgrimage are subject to significant changes (Norman and Mark 2011; Reader 2015; Gonzalez et al. 2019; Carbone et al. 2016; Kate and Hemel 2019), they require a special approach (spiritual, organizational, promotional, etc.) of church institutions in neoliberal times—the age of global religious pluralism. Most articles about mass events organized by churches or religious organizations consider individual mega-events, e.g., World Youth Day. There are publications presenting all aspects of this particular event (1985–2018), but they mostly fall into the category of popular writing (Muolo 2015; Sarniewicz and Anna 2016). There are, however, academic publications describing individual events, and one of the first studies on World Youth Day was conducted in Cologne in 2005 (Pfadenhauer 2010). Many papers were published after the next event, which was held in Sydney in 2008 (e.g., Webber 2008; Rymarz 2008; Singleton 2011; Mason

[1] Pilgrims from all over the world stayed in Rome (1986), Buenos Aires (1987), Santiago de Compostela (1989), Częstochowa (1991), Denver (1993), Manila (1995), Paris (1997), Rome (2000), Toronto (2002), Cologne (2005), Sydney (2008), Madrid (2011), Rio de Janeiro (2013), Krakow (2016) and Panama (2019).

2012; Halter 2013). WYD in Rio de Janeiro was described among others by de França et al. (2014) and Gonzalez et al. (2019), while the latest WYD, held in Krakow, was widely discussed by Polish researchers in individual articles and detailed reports (Jackowski et al. 2016; Kozak and Pęziński 2016; Nowotny 2016; Bogacz-Wojtanowska et al. 2016; Borkowski 2017; Seweryn 2018). It is uncommon, however, to find publications on the initial stage of World Youth Day—the Days in the Dioceses—which has an influence on the promotion of individual regions of the host country. Two works on this topic were published by Bik (2016) and Bik and Stasiak (2017).

3. Research Aims, Methods and Source Materials

The most important aims set by the authors were to analyze the promotional and organizational activities connected with a global religious event, describe the participants of World Youth Day in the Archdiocese of Lodz, as well as establish the impacts of the event on the region, especially economic and promotional.

The authors posed three research questions:

Q1—How are contemporary religious mega-events organized?

Q2—Who were the participants of the World Youth Day in the Archdiocese of Łódź and what did they do in Łódź in addition to participating in religious events?

Q3—How did the pilgrims' stay impact on the Łódź region?

Depending on the aim, the authors used different research methods and source materials, including the following:

- Participant observation—one of the co-authors was a volunteer at the WYD Diocesan Centre of the Archdiocese of Lodz for 30 months (she was responsible for contacts with foreign groups and for pilgrims' accommodation);
- A questionnaire—the offices of the four festivals in Lodz provided documents, organizational information and above all databases, including detailed statistics on nearly all pilgrims (9952 out of the 10,173 officially registered). The authors also used data obtained from the Central Statistical Office and the WYD Organizing Committee in Krakow;
- A pilot survey conducted at the end of 2016 (after the pilgrims returned home)—an invitation was mailed to over 1500 people; the questionnaire posted on the internet was filled in by 258 respondents (121 Poles and 137 foreigners from 11 countries); the questionnaire (in five languages: Polish, English, French, Italian and Spanish) contained 35 questions, e.g., reasons for participating in WYD, sources of information about Lodz, the character of their stay, ways of spending free time, opinions and assessment, as well as personal data;
- An analysis of media reports concerning the Days in the Dioceses which appeared on 15–25 July 2016 on social media (YouTube), on official WYD websites (Kraków 2016 2019; Łódź Piotrowa 2019; Archidiecezja Łódzka 2019; Magis Poland 2016 2019; Paradise in the City 2019), on regional TV and in the press. The analysis included over 100 video broadcasts and films, as well as 30 articles containing 84 statements by foreigners and 63 by Poles.

The material presented in the article is only a fraction of the results of research conducted by the authors as a part of a broader study of tourism in the Lodz region.

4. World Youth Day 2016 in the Archdiocese of Lodz

The Archdiocese of Lodz is one of 41 Roman Catholic Church dioceses in Poland. It covers 5200 km^2 and is situated in the very center of the country, but its borders are not congruent with the borders of Lodz Province (it covers ca 30% of the territory and includes 60% of the region's population). The history of the Archdiocese of Lodz is relatively short. The rapid development of the textile industry in Lodz in the 19th century, the huge inflow of people to the city, as well as the new political situation after World War I were the reasons why, in 1920, Pope Benedict XV created a new diocese with a seat in Lodz. After 72 years (25 March 1992), Pope John Paul II raised it to the rank of archdiocese. Since 2004, together with Łowicz diocese, it has constituted the metropolitan archdiocese of Lodz (one of 14 church provinces on the territory of Poland). Currently, it has about 1,430,000 believers. Its area is divided into 219 parishes (27 deaneries), employing 570 priests (www.archidiecezja.lodz.pl). The best-known figures connected with Lodz include St Maximilian Kolbe[2] and St. Faustina Kowalska[3].

World Youth Day in 2016 was one of the most important events in the short history of the Archdiocese of Lodz. The Days in the Dioceses featured four festivals, prepared by four separate organizers, cooperating with one another:

1. Peter's Boat—the main festival prepared by the WYD Diocesan Centre of the Archdiocese of Lodz attracted 2161 participants from 16 countries;
2. The Claretian Family Youth World Meeting—a meeting of the communities and milieus which are active in Claretian parishes all over the world (formally independent, but often organizationally linked with "Peter's Boat"); 231 participants from 12 countries;
3. MAGIS 2016[4]—coordinated by Jesuits, a meeting of young people from all over the world, united by Ignatian spirituality[5]. In Lodz, only the first stage of the MAGIS program took place; after three days, about 2500 participants (representing nearly 50 countries) went back to nearly 100 locations in Poland, the Czech Republic, Lithuania and Slovakia, to take part in other classes and workshops;
4. Paradise in the City—the largest (5345 participants from 48 countries) festival, prepared by the French Chemin Neuf Community[6].

Each festival had its own, individual program consisting of various religious events (services, meditations, vigils), as well as social (meetings in international groups), tourist (sightseeing, active hiking, e.g., kayaking, biking), cultural (concerts) and sports events (football, volleyball). The Days in the Dioceses closed with a farewell holy mass, celebrated in the Atlas Arena sports hall for the participants of all festivals, attended by about 13,000 inhabitants and pilgrims.

It is worth mentioning that the pilgrims staying in the Archdiocese of Lodz travelled over the whole Lodz region, and Days in the Dioceses events were also held in other dioceses of Lodz Province (Figure 1). Visits to Świnice Warckie, where St. Faustina was born, were particularly significant.

---

2 Maksymilian Maria Kolbe (born Rajmund Kolbe, 1894–1941)—Polish Conventual Franciscan, missionary, martyr, saint of the Roman Catholic Church. Founder of the most numerous Catholic monastery in the world in Niepokalanów (762 people in 1939). He voluntarily went to death by starvation in exchange for another prisoner of the Nazi concentration camp—KL Auschwitz. The first Polish martyr of the Second World War (beatification in 1971, canonization by Pope John Paul II in 1982).

3 Maria Faustyna Kowalska (born Helena Kowalska, popularly spelled Faustina, 1905–1938)—a saint of the Roman Catholic Church, a nun from the Congregation of the Sisters of Our Lady of Mercy, a mystic, stigmatist and visionary. A proclaimer of the cult of Divine Mercy. The author of the Diary, in which she described her spiritual and mystical experiences (a Polish book most often translated into other languages). Beatification in 1993, canonization in 2000 (both by Pope John Paul II). In 2005, she was announced the patron of Łódź.

4 The Latin word *magis* means: *more, better, fuller*.

5 Ignatian spirituality (Jesuit spirituality)—a method of shaping and leading the inner and outer life of a Christian, based on three spiritual exercises proposed by Ignacy Loyola (founder of the Jesuit order): "Principle and foundation", "Call of the King" and "Contemplation to obtain love".

6 The Chemin Neuf Community was formed in Lyon, in 1973. Its mission is evangelization and Christian spiritual formation in mature faith. It operates in about 30 countries all over the world and includes nearly 2000 members, including about 100 in Poland (Chemin Neuf Community 2019).

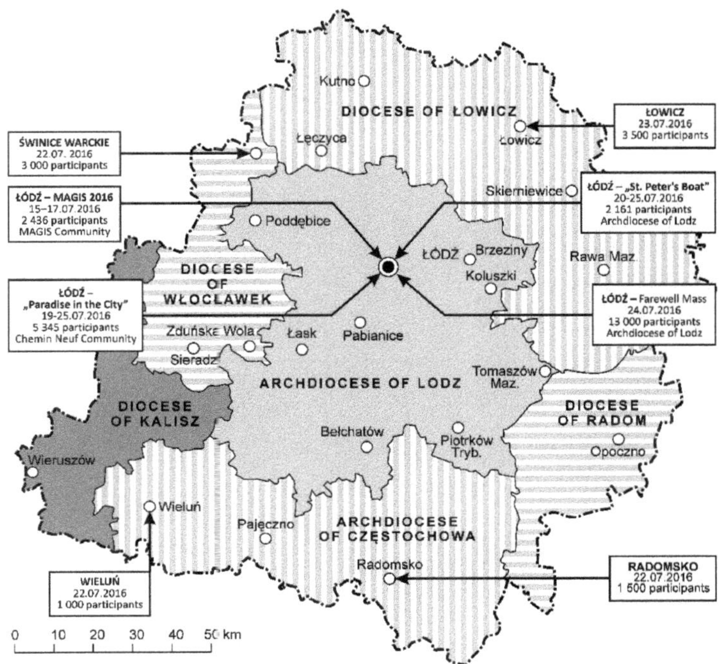

Figure 1. Major World Youth Day (WYD) events in 2016, in Lodz Province (Source: authors).

5. Results

5.1. The Organization of the Days in the Dioceses in the Archdiocese of Lodz

The WYD Organizing Committee of the Archdiocese of Lodz was officially established on 2 February 2015. It consisted of 16 sections (Figure 2), forming the following four sectors:

1. Administrative—dealing with all kinds of office work (WYD Office sections: Invitations, Information and Accommodation, Contracts and Finance, Translations);
2. Event preparation—caring for the pilgrims' basic needs and safety, as well as the necessary infrastructure (sections: Transport and Logistics, Catering, Maintenance, Medical Care and Security);
3. The form of the event—taking responsibility for the program, the course of the event and visuals (including the media) (sections: Liturgy, Music, Exhibition, Media);
4. Cooperation—coordinating volunteers' work and contacts with Catholic organizations and movements, representatives of other denominations, as well as representatives of the state, local authorities and security institutions.

Almost all of these sectors were headed by a Chairman, Vice-chairman, a Director of Operations and a secretarial office.

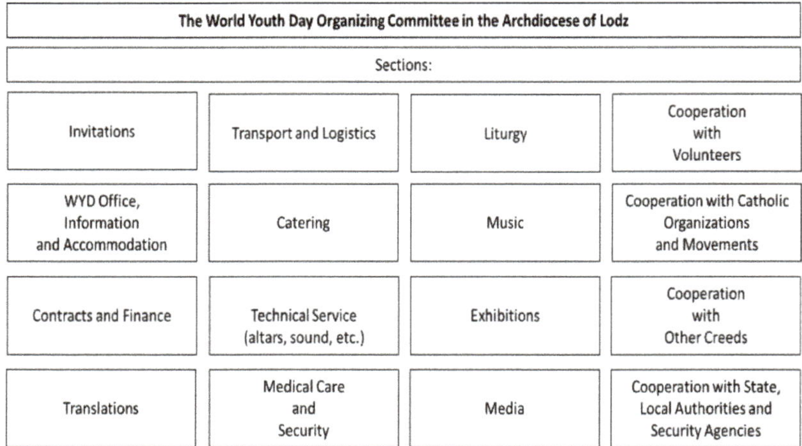

Figure 2. Structure of the WYD Organizing Committee of the Archdiocese of Lodz, 2016 (Source: authors).

For the duration of WYD, the area of the Archdiocese of Lodz was divided into special spatial units consisting of several adjacent deaneries. There were 12 of them, including four in the city of Lodz. This organization was to facilitate the preparation and management of the event outdoors. Each area was headed by a regional coordinator, directly subject to the WYD Diocesan Centre of the Archdiocese of Lodz.

As individual dioceses had to attract pilgrims (Krakow Committee did not suggest, let alone allocate, places of stay in Poland before WYD), it was necessary to introduce a number of promotional activities that were supposed to encourage young people from all over the world to come to Lodz. It was not easy to do because the Archdiocese of Lodz is relatively new, and the city and the region are not world famous or considered to be particularly attractive in Poland. It was necessary to make the area stand out from other places. The Lodz region was presented, first of all, as the place of St. Faustina Kowalska's first revelations and the place where St Maximilian Maria Kolbe had lived as a child. The promotional messages often stressed the past multi-culturality of Lodz and its industrial character, as well as its closeness to Krakow and Warsaw (Lodz is located in the very center of Poland).

As regards members of religious orders (Claretians and Jesuits), or the international Chemin Neuf Community, attracting them largely focused on encouraging people involved in pastoral activity around the world to come to Lodz. It was different in the case of the Peter's Boat festival, where initially there were no organizers or potential participants. It was necessary to plan and run a professional promotional campaign on an international scale, which included:

- Developing a corporate identity (a symbolic name—Peter's Boat[7], logo, colors: red, blue, yellow; hymn—"Blessed be the Merciful");
- Starting an event website (www.lodzpiotrowa.com) and a Facebook profile (WYD/SDM Lodz Piotrowa);
- Developing the Archdiocese of Lodz Press Office which was responsible for contacts with foreign journalists, official reports (over 20 video reports), promotional videos (8), interviews, etc., as well as for organizing regular press conferences (4);

[7] For the duration of WYD, dioceses in Poland took on biblical names. "Peter's Boat" refers directly to the meaning of the Polish name of the city (*łódź* = boat), as well as to the pilgrimage of John Paul II to Lodz in 1987, when the expression "Peter's Boat" was used for the first time.

- Making and popularizing special video reports from the preparations for the event (over 25);
- Designing and printing promotional materials (leaflets, brochures, etc.) in different languages (Polish, English, French, Italian, Spanish);
- Producing several thousand promotional items advertising Peter's Boat (ballpoint pens, mugs, bags, buttons, T-shirts, bracelets, CDs, key rings, etc.);
- Running training courses for volunteers in the media section (Youth Academy of Journalism and Journalism Workshops, organized by the Polish WYD Committee[8]);
- Direct promotion of the event, e.g., during the Taize European Meeting of Young Adults, the European Meeting of the Carmelite Youth in Avila and the International Convention of WYD Delegates in Wieliczka where each of the dioceses could put up its stand;
- Sending out the offer (program and promotional materials) to dioceses in other countries;
- Promoting WYD among the inhabitants of the region (e.g., advertisements on public city transport vehicles (MPK) and Lodz Agglomeration trains).

The activities proved effective. The Archdiocese of Lodz was visited by about 10,173 pilgrims, which can be considered a success, with the fourth best result in the country after the Archdioceses of Warsaw, Wroclaw and Katowice (ca. 12,000 in each).

In order to make the process of registering for the event smoother, special on-line registration forms were prepared (separate for Poles and foreigners and in different language versions). After registering a group, it was contacted only by the registration coordinator. Due to the risk of terrorist attacks, all potential participants of the event were verified by the Home Security Agency (pol. Agencja Bezpieczeństwa Wewnętrznego—ABW). The pilgrims were distributed over the area of the whole diocese, according to earlier offers from host families. The groups were picked up on arrival from a place agreed earlier (airport, train station) and transported to the pilgrims' registration point, often escorted by the police.

Such a huge logistic operation required the involvement of over 1000 volunteers. Earlier, they had been thoroughly checked by the police and the Home Security Agency. They also went through extensive training (e.g., on first aid, mass event protection, and directing traffic at railway stations) conducted by appropriate services (the police, emergency services, ABW, Railway Police, Fire Brigade). The volunteers were allocated specific regions and tasks. Before the last mass celebrated in the sports hall (Atlas Arena) the volunteers checked bags, vehicles and tickets at the entrance gates (in cooperation with the police), issued name badges, directed the traffic, provided first aid, worked on stalls, and distributed water and food.

To sum up, the professionalism of organizing the World Youth Day can be seen in the following:

- The formalized and developed organizational and spatial structure of the event;
- Various modern activities promoting the events, both on a national and international scale;
- An online system for registering groups of pilgrims in different language versions;
- Services provided by a group of over 1000 trained volunteers;
- Effective realization of several dozen small events (religious, cultural, tourist, recreational and sport-oriented).

5.2. *The Participants of World Youth Day in the Archdiocese of Lodz*

The Days in Dioceses in the Archdiocese of Lodz were attended by 10,173 people from nearly 80 countries around the world. The majority (over 77%) were Europeans (Figure 3). The remaining

[8] The WYD Committee, appointed by the Archbishop of Krakow, was responsible for the organization of the main celebrations of the 2016 World Youth Day. A total of 120 employees and 50 long-term volunteers worked there. Events within the Days in Dioceses were prepared by diocesan organizing committees. On 2 February 2015, Archbishop Marek Jędraszewski issued a special decree establishing the WYD Organizing Committee in the Archdiocese of Łódź.

continents were represented by more or less equal numbers of pilgrims: Asia—5.65%, Africa—5.20%, South and Central America—4.67%, North America—4.54%. They included representatives of some exotic (not popularly known to Poles) countries, such as Benin, Burkina Faso, Chad, Guyana, Guadalupe, Haiti, Congo, Madagascar, Martinique, Mauritius, New Caledonia, the Philippines, Rwanda, Reunion, the Seychelles or Zimbabwe. Australia and Oceania (the farthest from Poland countries) were represented by the smallest number of participants (under 0.5%). However, some pilgrims arrived in Lodz from as far as New Caledonia in the Pacific Ocean, which is located more than 15,500 km away from Poland.

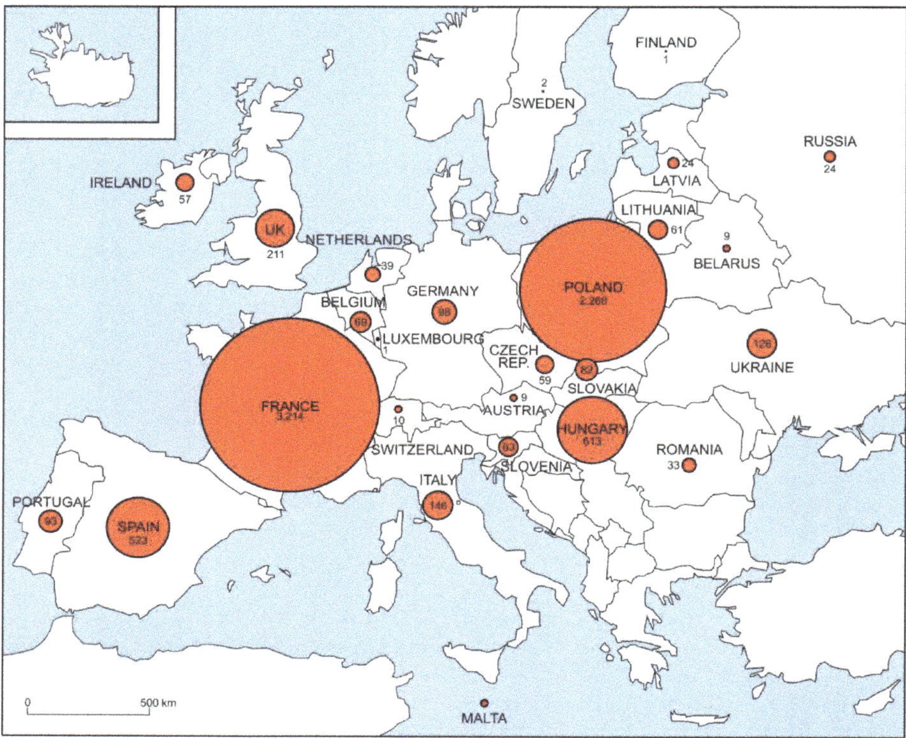

Figure 3. Origins of the European participants of the Days in the Dioceses in the Archdiocese of Lodz (Source: authors, based on the data provided by the organizers).

Paradoxically, the nation which was most numerously represented at the World Youth Day (WYD) in the Archdiocese of Lodz were not Poles (2268), but the French (3214), which is easy to explain. The organizer of one of four festivals was the French community Chemin Neuf, which very efficiently recruited pilgrims, not only in France, but in many other countries as well. As a result, the "Paradise in the City" festival attracted the greatest number of pilgrims (5300) from 48 countries (52.5% of the participants of all events, 60% of all the participating countries).

Besides the French and Poles, other nationalities were represented by far smaller numbers: Hungarians—by 613 people, Spaniards—523, Canadians—280, the British—211, and Americans—by 182. Most countries were represented by only a few or several pilgrims. We can observe here a clear relationship: the higher the percentage of Catholics in a given country (>50%) and the shorter the distance from Poland (<1500 km), the greater the number of pilgrims arriving for WYD festivities.

An additional factor encouraging people to travel to Lodz was the dynamic activity of Chemin Neuf or religious orders (Claretians, Jesuits) in a given country.

It is noteworthy that only the French, Poles, Spaniards, Canadians and Brazilians participated in all four festivals held in Lodz. Representatives of most nations took part in one festival (maximum two).

Based on the data obtained from the organizers of three festivals, it was possible to define the exact age and sex structure of the participants. A considerable majority of the pilgrims were people aged 15–30 (mostly in the 20–24 and 15–19 ranges), as the event is, by definition, directed towards young people. However, because the age limit is not observed too rigorously at the WYD, it was attended by both younger (under 10) and older (over 70) pilgrims. In the first case, they were usually children accompanying their parents. The other group consisted of priests, bishops, nuns and youth carers. They also included many participants—older by birth certificate but young at heart. The majority of people attending the Days in the Archdiocese of Lodz were women, who outnumbered men by 14%. What is interesting is that men did not form the majority in any of the age groups.

Following the WYD organization arrangement, most foreign pilgrims were accommodated with families in Lodz and other destinations in the region (59% of the people participating in Peter's Boat, Paradise in the City and Claretians' festivals). The remaining 41% stayed in schools and student hostels. There were many people willing to take the pilgrims into their own homes, but the festival organizers offered student's hostels (Technical University and University of Lodz) to the participants for their convenience, with the main attractions and events nearby.

More detailed information was provided by the survey that was conducted among the pilgrims. The relatively small sample (n = 246) cannot be regarded as fully representative of the whole group. The conclusions must be then approached carefully. The results that were obtained, however, give a general idea of the motivations, behaviour and opinions of the WYD participants. It was assumed—and later fully confirmed—that the Days in the Diocese were attended by two separate groups: Poles (49.2%) and foreigners (50.8%). Therefore, the description of WYD participants will be a comparative analysis of these two populations.

A vast majority of the respondents had participated in World Youth Day for the first time (nearly 91% of Poles and over 2/3 of the foreigners). Only 9.1% of Poles and nearly 1/3 of the foreigners stated that they had already taken part in a young people's meeting with the Pope before. What is important is that it was not confirmed that earlier experience of the event made the respondents more critical in their assessment and opinions.

Foreign pilgrims usually stayed at the Archdiocese of Lodz for 4–5 days (52.6%) or 6–7 days (36.3%). Shorter or longer stays were infrequent. Next, everybody travelled to Krakow to take part in the main WYD festivities. Thus, the foreigners spent a total of 10–14 days in Poland. The participation of Poles was slightly different. Due to their involvement in organizing the event, their stays were visibly longer. Nearly 1/3 of the respondents spent over a week in Lodz, 27.3%—4–5 days and 23.1%—6–7 days. There were also Polish pilgrims who appeared only at individual events for one (9%) or 2–3 days (8%). Including one week in Krakow, the average stay of Poles at World Youth Day can be estimated at 2–3 weeks.

The most popular means of transport for foreigners were the plane and the coach. Approximately half of the travelers (mostly from other continents) came to Poland by plane, and the other half (inhabitants of European countries) came by hired buses and coaches. Other means of public transport (e.g., trains, regular buses), as well as private cars, were statistically insignificant. The trips were usually prepared by various church institutions (parishes, religious communities, special diocesan bodies, orders, etc.—jointly over 70%), or travel agencies from the pilgrims' countries, specializing in religious tourism (11%).

The institutional organizer of the trip decided not only about the means of transport to be used, but also about a person's companions during the stay in Poland. Therefore, the foreigners usually arrived in Lodz in organized groups, together with other members of their parish (over 42%) or another religious community (23.7%), possibly in a group of friends (28.1%). Being accompanied by family members, partners or traveling alone was reported by very few people. Poles' responses were quite different. Over half of them (52.9%) declared that they were taking part in the WYD events together with their friends and people they knew. Every third respondent was a member of a larger group (a parish community or another religious organization). Every tenth respondent arrived at WYD alone, with no company.

It was particularly interesting to analyze the answers to the question about the motivations and purposes of participating in World Youth Day. They were divided into two groups: the main reasons—religious (Figure 4) and the remaining ones—non-religious (Figure 5), presenting a range of several possible variants of responses. The respondents also had an opportunity to make a free comment to this open question.

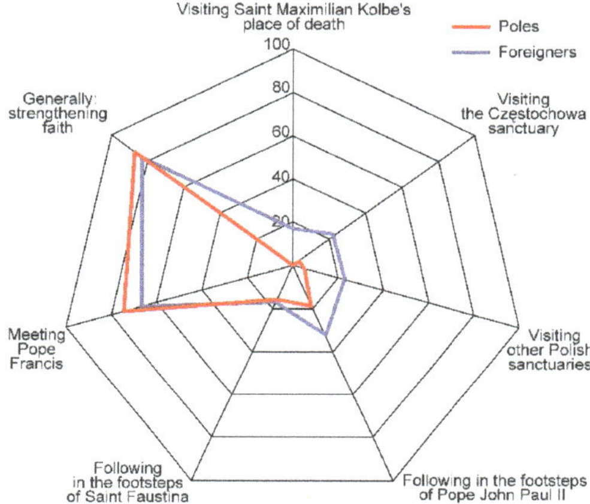

Figure 4. Religious purposes of respondents' participation in World Youth Day. (Source: authors, based on a survey).

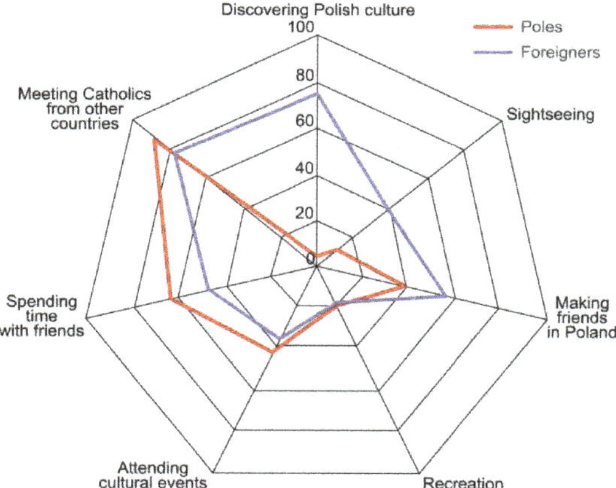

Figure 5. Non-religious purposes of respondents' participation in World Youth Day. (Source: authors, based on a survey).

As regards the religious-spiritual purposes of attending WYD, two were clearly predominant: "the overall strengthening of faith" (over 80% of the respondents) and "the meeting with Pope Francis" (ca. 70%). Those responses were most common among Poles and foreigners alike (Figure 4), with a few percent of Poles more (as regards their desire to see the head of the Catholic Church, the difference was 8%—74% compared to 66%). As for the foreigners, the "religious-tourist" purposes turned out to be more important. They, much more often than Poles, wanted to visit places related to Pope John Paul II (33% and 19%, respectively). Częstochowa and other Polish sanctuaries (22% and 3%, 23% and 5%, respectively), as well as KL Auschwitz, where St. Maximilian Kolbe had died (16% and under 1%, respectively). A similar proportion of Polish and foreign respondents (16–17%) planned a trip following in the footsteps of St. Faustina.

Even bigger differences could be noticed as regards the non-religious purposes (Figure 5). Poles tended to prioritize "social" motivations: meeting Catholics from other countries (over 88%), spending time among friends (63%), as well as attending cultural events (42%). They were also important to the foreigners but indicated by them much more rarely (from a few to several percent). The only exception was the desire to make new friends in Poland. The hope to meet new, interesting people was expressed by 56% of foreign and 39% of Polish World Youth Day participants. Many more foreign than Polish pilgrims indicated "tourist" purposes: discovering Polish culture (76%), sightseeing (39%), as well as visiting museums, discovering Poland's natural assets and practising active tourism (several percent each). Only a small proportion of Poles were interested in these forms of activity. The two groups were in complete agreement as regarded the recreational purpose. Every fifth respondent, regardless of their origin, hoped to relax and rest.

The respondents described their motivations in more detail in the open question. Naturally, the answers were extremely varied. After being put in order and aggregated, they can be divided into the following groups of motivations:

- Religious-spiritual, e.g., deepening the faith, sharing faith, meeting with God, meeting the head of the Church, pilgrimages to sacred places, recollections, spiritual development, service to the Church, experiencing the community (unity) of the Church;
- Cognitive-cultural, e.g., being curious of another country, other people, a different culture, the desire to travel, visit new places, practice English (mostly Poles), visiting friends and relatives (occasionally the foreigners);
- Socializing, e.g., striking new international friendships, spending time among peers, meeting people of similar views and professing the same values;
- Charity and "patriotic" motivations—those regarded only Poles, who often explained their volunteering with a strong, inner need to give something to others, to serve a fellow man, as well as with the feeling of pride with Poland and the wish to show their country, city and Church to foreigners (see too: Seweryn 2018).

Pilgrims from abroad, with a few exceptions, had never been to Lodz or the Lodz region before. They did not have any family bonds with Poland either. Nearly half of the respondents had not known anything about Lodz and its region before arrival (the stay was organized by church institutions, which did not require the participants' involvement). Those who searched for information about the place of their stay looked it up on the Internet (37%), asked their families and friends (17%), looked for books and tourist guidebooks (13%), and every tenth visited a tourist information point on the spot, in Poland. In Lodz, the most frequently used information centres were located in the historical centre of Lodz (Manufaktura, Piotrkowska Street) and at the main transport nodes (Łódź Kaliska and Łódź Fabryczna train stations, W. Reymont Airport). Local guides, both professional and amateur (Polish volunteers and carers/supervisors), were an important source of information about the region. Nearly half of the foreigners admitted having used their services.

The pilgrims' schedule for their stay in the Archdiocese of Lodz was filled primarily with all kinds of religious-spiritual events, not leaving much time for tourism or recreation. Therefore, visiting the region was usually limited to walking around the town and making short trips to nearby destinations.

The most frequently indicated place in Lodz, visited by the respondents during the Days in the Dioceses, was Atlas Arena—a sports hall, where the final mass, closing the whole event was held

(84% of Poles and 75% of foreigners—cf. Figure 6). Other places that were frequently mentioned by the respondents included St. Stanislaw Kostka Cathedral, Poniatowskiego Park, "Bombonierka", Expo Hall and the City Centre of Sport and Recreation, Słowackiego Park (Venice) and St. Faustina's house, i.e., sites located in the southern part of the city centre, where the pilgrims' life was bustling.

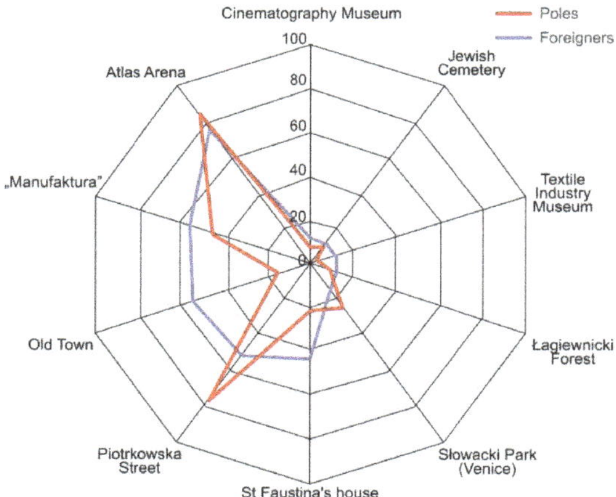

Figure 6. Lodz attractions visited by respondents (Source: authors, based on a survey).

Apart from taking part in religious events, the WYD participants could freely walk around the city. They were mostly interested in the historical centre of Lodz. Over half of the foreigners visited Manufaktura[9], Piotrkowska Street[10] and the Old Town[11]. Poles' preferences were slightly different. 3/4 went to Piotrkowska Street, half of them—to Manufaktura, and only 16%—to the Old Town. The differences in responses, however, may result from the misunderstanding of the term "Old Town". The foreigners quite often considered the representational Piotrkowska Street to be a fragment of the oldest district of the city, which is not true. It is a fact, however, that all these three spaces have a historical-commercial-recreational character. Other tourist-recreational attractions of Lodz were far less interesting for the respondents. Every 8–9th foreigner visited Łagiewnicki Forest, the Museum of Textile Industry, the Museum of Cinematography and the Jewish Cemetery. Poles appeared there less often, but they additionally visited Księży Młyn and the Old Cemetery in Ogrodowa Street (10% each). Those two places were rather omitted by the foreign visitors. Only few people went to the Museum of Art, Palm House, "Fala" Aquapark or the Open-Air Museum of Wooden Architecture.

The respondents were also asked which event of the Days in Dioceses had impressed them most. Free and very emotional responses, given by Poles and foreigners alike, clearly pointed to the following: (1) the Dismissal Mass and the closing concert in Atlas Arena, (2) the Way of the Cross, led along the streets of the city, and (3) other, minor events (concerts, meetings, prayers), which took place during the festivals or in the parishes receiving the pilgrims.

The respondents also had the opportunity to evaluate different aspects of the Days in Dioceses (1 point = very bad to 5 points = very good). Both the events and the stay in the Archdiocese of Lodz itself were evaluated very highly by Poles and foreigners alike (Figure 7). The elements appreciated the

---

9 A shopping and recreation center, located on the premises of a former, revitalized 19th century textile factory, owned by Israel Poznański, one of the largest in the city.

10 The main street of Lodz, over 4 km long—a historical route and the central axis of the developing city in the 19th century. Today, the northern section of the street (2 km) is a representational promenade and the center of city life.

11 The area of the medieval city surrounding the Old Market (north of Piotrkowska Street and west of Manufaktura). In 1945–1944, the Old Town was situated within the premises of the Litzmannstadt Ghetto, created by Germans.

highest (over 80% were rated as good and very good) included the following: the atmosphere during the festivities (92%!), the kindness of the inhabitants, security and hospitality (Figure 8). The last of the above-mentioned categories reached the highest arithmetical mean (4.83 among Poles and 4.89 among foreigners). There were, however, visible differences between the evaluation by Polish and foreign respondents. The Poles rated cultural and sports events accompanying the WYD higher, while the foreigners—the accommodation, tourist information and the cleanliness of Lodz streets.

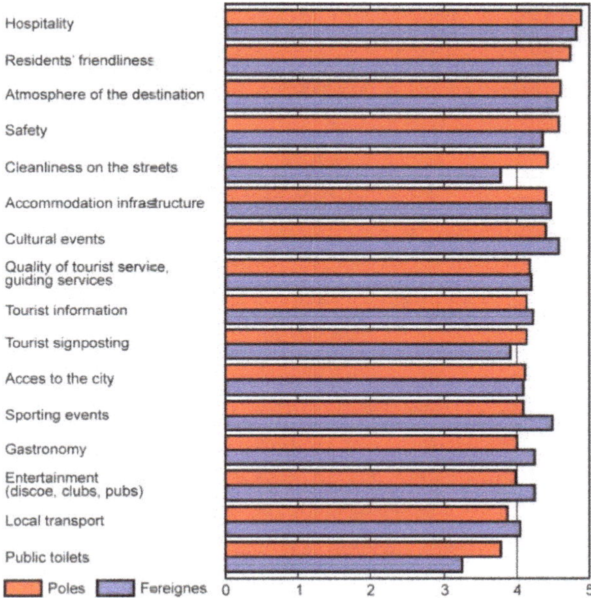

Figure 7. Mean ratings of different aspects of WYD in the Archdiocese of Lodz (1—very bad, 5—very good) (Source: authors, based on a survey).

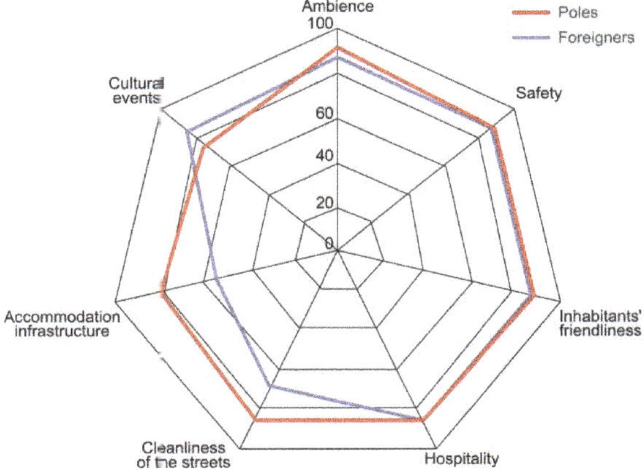

Figure 8. Top-rated aspects of WYD in the Archdiocese of Lodz (the total of "good" and "very good") (Source: authors, based on a survey).

The last of the above-mentioned aspects is particularly interesting. The foreigners praised the cleanliness of the city streets (84% of positive ratings, the mean: 4.42), while the Poles were more careful when giving good and very good marks in this category (68%, mean: 3.78). As much as 11.5% thought that the level of cleanliness in the public spaces of Lodz should be a cause for embarrassment (it was rated as "very bad" and "bad"). This opinion was shared by 10 times fewer foreigners (only 1.5%).

What differed the respondents most were probably the negative opinions (Figure 9). The Poles first of all criticized public toilets (16.5% were rated as very bad and bad, the mean: 3.26), the cleanliness of streets and tourist graphic identification (6.6%; 3.91). The foreigners were definitely more critical of the local transport (9.6% of negative opinions, mean: 3.87) and access to the city (6.7%; 4.11), as well as—which may seem a little surprising—the standard of gastronomic services (low for 8.1% of respondents, despite the mean of 4.00). In general, however, the visitors were much less judgemental than Poles.

Figure 9. Lowest rated aspects of WYD in the Archdiocese of Lodz (the total of "bad" and "very bad") (Source: authors, based on a survey).

Naturally, all the evaluations should be approached with extreme caution, as they are heavily subjective (lack of clear assessment criteria, various points of reference when evaluating, for instance, the standard of living in one's own country), as well as sensitive to external factors (the context of evaluation, personal experience during WYD, courteousness, the feeling of gratitude towards the hosts, etc.). Nevertheless, the general impression of the Days in Dioceses must have been very positive, as over 90% of the respondents (90.9% of Poles and 91.9% of foreigners) declared that they would recommend Lodz and the Lodz region to their families and friends. A great majority of the respondents also claimed that they would be happy to take part in the next World Youth Day (80% of foreigners and 68% of Poles). These responses confirm the assumption that the event attracts young, active Catholics involved in the life of their religious community.

To sum up, based on the survey results, it can be concluded that, at the diocese level, WYD was attended by two main groups: Poles (nearly 1/4) and foreigners (3/4). The former usually came from Lodz and the Lodz region, played the role of hosts, often worked as volunteers helping to organize the event, spontaneously took care of the visitors, and took on the role of the local guides and animators of the guests' free time. The latter, regardless of the country they had arrived from, behaved as guests, showed great interest in Poland, and were willing to take part in cultural, sports or tourist events. They can certainly be defined as pilgrim tourists or religious tourists. Despite the fact that the main, spiritual aim of the meeting was the same, the groups differed considerably, for example, as regards the motivations to participate in World Youth Day, religious or tourist-recreational behaviours, as well as evaluations and opinions. All of them, however, created a unique atmosphere of an exceptional holiday—a joyous meeting of the Catholic youth from all over the world.

5.3. Economic Outcomes

Expenditure levels during the Days in the Dioceses are extremely difficult to define, or even calculate, due to different forms of payment and varied charge rates, where factors such as the home country or the status of the participant have to be taken into consideration as well. Participating in the diocese stage on the premises of the Archdiocese of Lodz cost from EUR 12 (Polish volunteers used a subsidy from Lodz Curia) to EUR 220 (foreign participants of MAGIS 2016 and Paradise in the City festivals)[12]. It was the price of the pilgrim's package, which included accommodation, board, water to drink, transport at the place of stay and transport to Krakow to the main WYD celebrations, insurance and a set of promotional items. The packages enabled the pilgrims to satisfy all their basic needs. Foreigners additionally paid for transport to and from Poland. This cost depended on the distance and the means of transport they chose.

According to the information provided by the Church, the payment for participation in individual festivals was used for their organization, covering the cost of renting facilities, transport, provisions for the pilgrims, sound systems, lighting, decoration, creating promotional materials and name badges, supplying water, organizing additional attractions, etc. However, they were not sufficient to cover all the costs of organizing the event. Some costs were covered by the families receiving the pilgrims; they not only put them up, but also gave them breakfast, drinking water and other support (transport, organizing free time, etc.). Some funds were provided from the state budget or local government budgets (security, medical aid, etc.).

Considering the rates suggested by the Episcopacy of Poland[13], it can be estimated that the income generated by over 10,000 visitors in the Archdiocese of Lodz reached at least ca. EUR 300,000. On the other hand, assuming average charges (medium price range), the sum of EUR 950,000 is obtained. That money above all went to small, local firms (catering, sound management, lighting, transport, printing, etc.). Such enterprises certainly significantly benefited from providing services for the event, but region-wise, those were still rather small sums.

The respondents taking part in the survey were asked about their expenses during their stay in the Archdiocese of Lodz (apart from the pilgrim's package and transport to and from Poland). Nearly every tenth foreigner and every third Pole declared a lack of any additional expenses. The remaining mentioned small, spontaneous shopping (bottled drinking water, extra food, devotional items, souvenirs). The average foreign pilgrim's expenses were about EUR 53 and Poles spent about EUR 32. These sums are significantly smaller (2–3 times) than the expenses of the tourists visiting the city and region of Lodz[14] in 2016 (despite the fact that the pilgrims stayed longer).

Assuming that the respondents' declarations concerning their expenses reflect the shopping behaviors of the remaining WYD participants, we may try to calculate the overall amount of money spent in the region by the pilgrims (Bik and Stasiak 2017). This sum is not very large, amounting to ca. EUR 400,000. However, the participants of the Days in the Dioceses did not arrive in order to shop, but for quite different, non-commercial purposes. The majority of pilgrims were not interested in spending money at all. Only one third of the respondents admitted that they would be willing to spend more, above all on food, souvenirs and sightseeing.

5.4. Promotional Outcomes

One of the WYD ideas that was particularly alive during the Days in the Dioceses was developing new friendships and sharing faith by young people coming from various parts of the world (Jackowski et al. 2016). Pilgrims' learning about the culture and traditions of a place they visit is

---

12 These are costs for the entire duration of stay.

13 The Episcopacy of Poland imposed the minimum charge for all foreigners in all the dioceses (30 Euro), so that they could not compete for foreign pilgrims through a lowered price of the packages.

14 Respectively: EUR 102 and EUR 103 (Napierała 2016).

most strongly affected by their personal experience and relationships with the inhabitants. It is later reflected in the perception they have of the whole visited country. Thus, it was decided to examine the perception of Poland by pilgrims from abroad, as well as the perception of the pilgrims by Poles. For this purpose, the authors thoroughly analyzed the contents of the media messages posted on the internet (YouTube, the local TV and the Archdiocese of Lodz news), as well as in the regional press (five newspapers). Among numerous reports from WYD, only those were selected which included direct comments made by the participants. In general, it was possible to record the opinions of 84 foreigners (73 on YouTube, 11 in the press) from 27 countries, including France (19), Spain (8), Zimbabwe (7) and Ukraine (6). It is possible to observe a certain tendency: the reporters interviewed representatives of the most numerous groups of pilgrims, but also representatives of countries which are exotic to Poles (e.g., Haiti, Martinique, Sri Lanka, New Caledonia). They participated in the two largest festivals in Lodz (Peter's Boat, Paradise in the City). The interviewees usually spoke about their motivations and purposes for participating in WYD and coming to Lodz, as well as evaluating their stay in Poland.

The main motivation to come to WYD was the desire to meet Catholics from all over the world (21 statements), to deepen faith (13) and to feel unity with other members of the Catholic Church (11). This last motivation was particularly important for people coming from countries where Catholics make up only a small part of society. In the context of the survey results, it may be surprising that for only five people it was important to meet the Pope, and nobody mentioned St. Faustina or St. Maximilian Kolbe.

It is much more difficult to sort out and generalize the opinions of the foreign visitors in Poland. Five said, generally, that they simply liked it there. However, the majority of responses were very emotional, full of pathos and exaggerated statements (Poland is extraordinary, wonderful, beautiful, breath-taking, etc.). Some pilgrims stressed that they felt very well there ("I feel here at home, though I've just arrived and I'm very far from my country"). They appreciated the way they were welcomed, the program that had been prepared for them to have fun and pray at the same time ("I'm very happy to be here with you—Poles. I like the way we are received here and I'm really proud to be here").

There were also comments regarding Polish food, which was described as "delicious", "good" or "different", as well as the weather in Poland, which some people loved (e.g., an Irish woman) and others thought it was too cold or too variable. Some visitors were surprised with the amount of greenery in Lodz, as well as the number and size of churches. Generally speaking, foreigners staying in the Archdiocese of Lodz greatly approved of Poland.

The pilgrims also commented on Poles, who often received them under their own roofs. The hosts were described in a very positive way, as amiable, friendly and kind people. This is interesting as it contradicts a popular self-stereotype that Poles are reserved, depressed, dissatisfied and complaining about everything.

During World Youth Day, the old Polish saying "a guest in the home, God in the home" became alive again. In the Archdiocese of Lodz, nearly 3500 pilgrims could experience Polish hospitality, staying with families and seeing what their everyday life was like. The foreigners often stressed in the interviews how well they were received by Poles and how nice their hosts were. They were surprised by Polish hospitality, including the meals they were offered ("We liked the people in Lodz best. They cared for us for all these days. This care we have experienced is really wonderful. The food is also delicious. Most of us lived with Lodz families. They prepared everything. The tables were always full. We really felt like home").

Both the hosts and the guests realized that there was a language barrier between them. In order to make the communication easier, they used Google translator ("Even elderly people manage Google very well, so they explain, they found great solutions and this exchange really is very, very fruitful and so very happy"), and they talked using gestures ("Most of us communicate with the families using gestures. But it's wonderful").

To sum up, foreigners highly evaluated both Poland and Poles. Table 1 presents 25 words which were most frequently used by the foreigners. The word "very" comes first—it was used 63 times, usually in combination with "much", "important", "nice", "good", "nicely". Other frequent words included "Poland", "here" and "Lodz", i.e., words describing the locations where the events took place. Based on the words which were used by the foreigners more often than ten times, we might risk defining the Days in the Dioceses in Lodz as a "wonderful experience, a very good meeting to share God here".

Table 1. Twenty-five words occurring most frequently in foreigners' and Poles' statements.

No.	Foreigners		Poles	
	Word	**Occurrence Rate**	**Word**	**Occurrence Rate**
1.	very	63	very	55
2.	Poland	40	youth	30
3.	here	31	joy	22
4.	Lodz	27	preparations	20
5.	meet	25	pilgrim	19
6.	God, Christ	24	meet	17
7.	well	17	together, all	13
8.	together, all	15	song	12
9.	share	13	extraordinary	12
10.	an experience	12	important	12
11.	wonderful	10	God, Christ	9
12.	experience	10	language	8
13.	faith	9	wonderful	8
14.	nice	9	to experience	8
15.	world	8	dance	6
16.	happiness	8	prayer	6
17.	meeting	6	surprise	6
18.	joy	6	cheers	4
19.	nicely	6	fatigue	4
20.	to experience	6	talk	4
21.	interesting	6	wait	4
22.	satisfaction	4	involvement	4
23.	great	4	chance	3
24.	extraordinary	3	testimony	3
25.	Church	3	smile	2

Source: authors.

Poles' comments in the media underwent a similar analysis (the total of 63, including 55 video broadcasts on YouTube and eight press reports). The interviews concerned above all three principal issues: the program and organization of the Days in the Dioceses, the opinions of host families, and general expressions describing the pilgrims and WYD. Different forms of the word "prepare" were used 20 times, which points to the fact that the role of host on World Youth Day and the smooth running of the event were treated very seriously ("We were preparing with the group for the whole year, in a very organized way, so that our guests from Italy and Ukraine could sleep in our house, had something to eat and something to do. So that they did whatever they liked and when they liked. But they also had some free time. It was important that they learnt something about the city, our culture and us as Poles, as young people"). The amount of preparation was accompanied with traditional Polish complaining. Poles mostly complained of being very tired, but they also stressed their satisfaction.

The comments of the families receiving the pilgrims focused on the following:

- Motivations (a form of returning the hospitality of other people, e.g., during pilgrimages);
- Ways of overcoming the language barrier (usually with a smile and gestures, learning basic Polish words);
- Traditional Polish dishes (nearly 20 specialities were mentioned, including żurek, pierogi and home-made bread);
- Opinions about the guests ("smiling", "cheerful", "grateful", "helping").

Poles, similar to foreigners, most frequently used the word "very" (55 times, e.g., "we are very happy", "very much", "very important"). Other frequent words included: "youth" (30), "joy" (22), "preparations" (20), "pilgrim" (19) and "meet" (17). Two words were not used by foreign visitors: "fatigue" and "difficult". It could be said that the hosts focused on the organization and meeting pilgrims from all over the world, while the guests paid more attention to meeting God and other human beings.

Perhaps it was the meeting of these two different worlds (host's and visitor's) that was the essence of World Youth Day. The event is a tool for breaking cultural barriers, and the opening of one person to another. Patience, the will to communicate and a positive attitude on both sides, as well as the same values (faith in Christ) make them happy members of one, huge community (Kalinowska-Żeleźnik and Lusińska 2011).

6. Discussion

World Youth Day can be treated as a long-term marketing strategy of the Roman Catholic Church, aimed at refreshing its image and convincing young people that it is a resilient and modern religious institution (Pfadenhauer 2010). The attractive, diversified and colorful event is also a tool of new evangelization, serving to propagate the canons of faith, to keep the current believers and to win over new ones. In this context, WYD is treated as an example of Peter L. Berger's theory of counter-secularization (Lynch 2008), an innovation aimed at maintaining the continuity of Catholicism (Mandes and Sadłoń 2018) or a modern pilgrimage event of evangelistic nature (Rymarz 2008; Norman and Mark 2011).

It proves a challenge for the organizers to find new means of communication with young people. Thus, they widely use the media that are popular among the youth: television, Internet, social media (Hepp and Veronika 2009). WYD is held during the holidays, which, on the one hand, facilitate the organization of outdoor events and, on the other hand, make it possible to combine religious practices with leisure and traveling (Mandes and Sadłoń 2018). Tourism allows the young to achieve values that are important to them: to see new places and to meet new people. Gonzalez et al. (2019) emphasize the important role of home hospitality as representation of religion in action. Thanks to everyday direct contacts, pilgrims and the families that accommodate them become closer to one another, break down intercultural barriers (Kalinowska-Żeleźnik and Lusińska 2011), enter into cordial relations, but also mutually strengthen their faith. Visitors have the chance to learn what the real life of the inhabitants of the visited country looks like. The practice to host the participants of WYD free of charge is several decades older than the contemporary vogue for couch surfing.

Carefully prepared and elaborate program, which has been evolving for years, is of great importance for the popularity of WYD. It is a "hybrid event" (Pfadenhauer 2010) that combines traditional, conventional religious elements with modern festivals, often of wholly lay nature. Hence, huge religious events constitute a sui generis bridge between traditional religious culture and modern popular culture, enable transmission of religious content in a manner understandable for the contemporary man (Zduniak 2015). That is why, for instance, the culminant meeting with the Pope is surrounded with multiple cultural, sports, tourist and recreational attractions that cater for all tastes and meet different expectations of the participants (Bogacz-Wojtanowska et al. 2016; Jackowski et al. 2016; Kate and Hemel 2019).

Organized since 1997, the "Days in Dioceses" component is destined for new pilgrimage tourism: spiritual preparation for the main celebrations, but also relaxation, getting to know the host country (its attractions, culture, cuisine, etc.) and meetings with its inhabitants. For the young, it is a perfect opportunity to experience religion in a completely new, emotional way, to experience unity of opinions with peers from other countries and participate in diverse events of hybrid (religious and lay) nature (Rymarz 2008; Mandes and Sadłoń 2018). And, when the opportunity arises, to travel, sightsee, discover, rest, entertain themselves. All the more so because this manner of getting to know the world does not require high spending and gives one a sense of security. In this regard, the obtained results correspond with the results of numerous other researchers (Kalinowska-Żeleźnik and Lusińska 2011; Bogacz-Wojtanowska et al. 2016; Jackowski et al. 2016; Kozak and Pęziński 2016; Borkowski 2017).

It is also worth pointing out that the eventization of faith satisfies not only fundamental spiritual needs, but also other human needs: biological and social ones, the need for security, for respect and acceptance, for self-fulfilment (Zduniak 2015) as well as for the sense of uniqueness, festiveness, celebration of special moments in life.

7. Conclusions

On the basis of wide-ranging and varied studies, we may draw the following conclusions:

1. Contemporary religious mega-events are not much different from other mass events. The main difference is the theme. All of them are complicated logistic undertakings, which require professional preparation and realization, effective organization by a team consisting of specialists in a variety of areas, as well as an army of strongly involved volunteers and public service workers. The attractiveness of an event to pilgrims is determined by its religious theme, but also the promotion campaigns, visuals, accompanying events, media coverage, etc. These activities demonstrate the constantly growing professionalism of the organizers of religious events, including World Youth Day.
2. During World Youth Day, held on 15–24 July 2016, the Archdiocese of Lodz was visited by over 10,200 pilgrims from 80 different countries. They were mostly young people (15–29), secondary school and university students, who had never been to Poland before. The main purposes of their trip were religious and spiritual (strengthening faith, meeting Pope Francis, following in the footsteps of St. John Paul II, visiting Polish sanctuaries). However, non-religious reasons also played an important role (e.g., meeting Catholics from other countries, discovering Polish culture and making friends in Poland). Despite a varied religious program, the pilgrims found time for fun, rest and sightseeing together. These additional attractions significantly enriched the program of their stay, highlighted the joyous character of the event, and built and strengthened the feeling of community and the exceptionality of the time spent together. In this sense, we can talk about clear symptoms of the eventization of faith.
3. Due to the relatively hermetic character of the event (focusing on the spiritual dimension, the organization mostly by church bodies, pilgrims rarely using external gastronomic and accommodation services), its influence on the economy of Lodz and the Lodz region should be considered insignificant. Organizing the Days in the Dioceses brought in a total income of about EUR 1.2–1.3 million. The total cost of preparing the event and actual staging remains unknown. World Youth Day above all had a promotional and image-creating function. Thousands of people from all over the world, who had not even heard about Lodz before, left the city satisfied and became its "walking proponents". The positive message (recommendations) spread among family and friends and on the social media should result in a more positive image of Lodz and the Lodz region (or even the whole of Poland) and, consequently, increased tourism in the near future. In this context, the co-financing of the event by state and local authorities (free use of the police, fire brigades, ambulances etc.) should be considered a kind of promotional investment.

Author Contributions: Conceptualization, J.B. and A.S.; methodology, J.B. and A.S.; software, J.B. and A.S.; validation, J.B. and A.S.; formal analysis, J.B. and A.S.; investigation, J.B. and A.S.; resources, J.B. and A.S.; data curation, J.B. and A.S.; writing—original draft preparation, J.B. and A.S.; writing—review and editing, J.B. and A.S.; visualization, J.B. and A.S.; supervision, J.B. and A.S.; project administration, J.B. and A.S. All authors have read and agreed to the published version of the manuscript.

Funding: This research received no external funding.

Conflicts of Interest: The authors declare no conflict of interest.

References

Archidiecezja Łódzka [Archdiocese of Lodz]. 2019. Available online: www.archidiecezjalodzka.pl (accessed on 14 September 2019).

Bik, Joanna, and Andrzej Stasiak. 2017. Światowe Dni Młodzieży w archidiecezji łódzkiej [World Youth Day in the Archdiocese of Lodz]. In *Ruch Turystyczny w Łodzi i Województwie Łódzkim w 2016 Roku*. Edited by Bogdan Włodarczyk. Łódź: Instytut Geografii Miast i Turyzmu Uniwersytetu Łódzkiego, pp. 145–68.

Bik, Joanna. 2016. Światowe Dni Młodzieży—Radosne spotkanie na gruncie wiary [World Youth Day—A joyful meeting on the basis of faith]. In *Kultura i Turystyka—Sacrum i Profanum*. Edited by Justyna Mokras-Grabowska and Jolanta Latosińska. Łódź: Regionalna Organizacja Turystyczna Województwa Łódzkiego, pp. 65–81.

Bilska-Wodecka, Elżbieta, and Sołjan Izabela. 2014. Religijne wydarzenia masowe [mass events] na przykładzie spotkań młodzieży chrześcijańskiej [Religious mass events on the example of Christian youth meetings]. In *Kultura i Turystyka—W Kręgu Wydarzeń*. Edited by Beata Krakowiak and Andrzej Stasiak. Łódź: Regionalna Organizacja Turystyczna Województwa Łódzkiego, pp. 67–80.

Bogacz-Wojtanowska, Ewa, Gaweł Łukasz, and Góral Anna, eds. 2016. *Światowe Dni Młodzieży 2016 Jako Fenomen Społeczny, Kulturowy i Religijny [World Youth Day 2016 as a Social, Cultural and Religious Phenomenon]*. Kraków: Stowarzyszenie Gmin i Powiatów Małopolski.

Borkowski, Krzysztof, ed. 2017. *Ruch Turystyczny w Krakowie w 2016 r. Uczestnicy Światowych Dni Młodzieży w Krakowie w 2016 r. [Tourist Traffic in Krakow in 2016. Participants of the World Youth Day in Krakow in 2016]*. Kraków: Małopolska Organizacja Turystyczna.

Carbone, Fabio, Corinto Gian, and Malek Anahita. 2016. New Trends of Pilgrimage: Religion and Tourism, Authenticity and Innovation, Development and Intercultural Dialogue: Notes from the Diary of a Pilgrim of Santiago. *AIMS Geosciences* 2: 152–65.

Chemin Neuf Community. 2019. Available online: www.chemin-neuf.pl (accessed on 14 September 2019).

Cudny, Waldemar. 2016. *Festivalisation of Urban Spaces. Factors, Processes and Effects*. Cham: Springer.

de França, Tiago Cruz, Diogo Nolasco, Rafael Lage Tavares, José Orlando Gomes, and Paulo Victor R. de Carvalho. 2014. A Critical Insight of the Pope's Visit to Brazil for the World Youth Day: Resilience or Fragility? Paper presented at 11th International ISCRAM Conference, University Park, PA, USA, May 18–21; Edited by Starr Roxanne Hiltz, Mark S. Pfaff, Linda Plotnick and Patrick C. Shih. Pensylvania: University Park.

Foley, Malcolm, McGillivray David, and McPherson Gayle. 2012. *Event Policy: From Theory to Strategy*. London and New York: Routledge.

Gartner, Wiliam C., and Donald F. Holecek. 1983. Economic impact of an annual tourism industry exposition. *Annals of Tourism Research* 10: 199–212. [CrossRef]

Getz, Donald. 1991. *Festivals, Special Events, and Tourism*. New York: Van Nostrand Rheinhold.

Getz, Donald. 2008. Event tourism: Definition, evolution, and research. *Tourism Management* 29: 403–28. [CrossRef]

Gonzalez, Luciana Thais Villa, Mariz Cecília Loreto, and Zahra Anne. 2019. World Youth Day: Contemporaneous pilgrimage and hospitality. *Annals of Tourism Research* 76: 80–90. [CrossRef]

Hall, Colin Michael. 1989. The definition and analysis of hallmark tourist events. *GeoJournal* 19: 263–68. [CrossRef]

Halter, Nicholas. 2013. The Australian Catholic Church and the Public Sphere: World Youth Day 2008. *Journal of Religious History* 37: 261–82. [CrossRef]

Hepp, Andreas, and Krönert Veronika. 2009. Religious Media Events: The Catholic 'World Youth Day' as an Example for the Mediatisation and Individualisation of Religion. In *Media Events in a Global Age*. Edited by Nick Couldry, Andreas Hepp and Friedrich Krotz. London: Routledge, pp. 265–82.

Jackowski, Antoni, Sołjan Izabela, Bilska-Wodecka Elżbieta, Liro Justyna, Trojnar Maciej, and Kostrzewa Edyta. 2016. World Youth Day in Kraków in the light of experiences from around the world. In *World Youth Days. A Testimony to the Hope of Young People*. Edited by Józef Stala and Andrzej Porębski. Kraków: The Pontifical University of John Paul II in Kraków, pp. 133–57.

Kalinowska-Żeleźnik, Anna, and Anna Lusińska. 2011. Spotkanie jako narzędzie przełamywania barier międzykulturowych na przykładzie Światowych Dni Młodzieży [Meeting as a tool for overcoming cross-cultural barriers on World Youth Day example]. *Zeszyty Naukowe Uniwersytetu Szczecińskiego, 647, Ekonomiczne Problemy Usług* 65: 139–49.

Kate, ten Laurens, and Ernst van den Hemel. 2019. Tensions and Interactions between Religious, Cultural and National Imaginaries in Neoliberal Times. An Introduction. *Interdisciplinary Journal for Religion and Transformation in Contemporary Society* 5: 259–81. [CrossRef]

Kozak, Anna, and Piotr Pęziński. 2016. *World Youth Day Krakow 2016. Selected Research Results*. Warszawa: Narodowe Centrum Kultury.

Kraków 2016. 2019. Available online: www.krakow2016.pl (accessed on 14 September 2019).

Łódź Piotrowa [Boat of St. Peter]. 2019. Available online: www.lodzpiotrowa.com (accessed on 14 September 2019).

Lynch, Andrew. 2008. Social theory, theology, secularization and World Youth Day. *New Zealand Sociology* 23: 34.

Magis Poland 2016. 2019. Available online: http://magis2016.org/pl (accessed on 14 September 2019).

Mandes, Sławomir, and Wojciech Sadłoń. 2018. Religion in a Globalized Culture: Institutional Innovation and Continuity of Catholicism—The Case of World Youth Day. In *Annual Review of the Sociology of Religion*. Leiden: Brill, vol. 9, pp. 202–21.

Mason, Michael. 2012. *Implication of Previous Research for World Youth Day Sydney 2008*. Fitzroy: Australian Catholic University.

Muolo, Mimmo. 2015. *Pokolenie Światowych Dni Młodzieży [Generation of World Youth Days']*. Kraków: Wydawnictwo św. Stanisława BM.

Napierała, Tomasz. 2016. Wydatki odwiedzających Łódź i region łódzki w świetle badań ankietowych [Expenses by visitors to Łódź and the Łódź region in the light of the survey]. In *Ruch Turystyczny w Łodzi i Województwie Łódzkim w 2016 Roku*. Edited by Bogdan Włodarczyk. Łódź: Instytut Geografii Miast i Turyzmu Uniwersytetu Łódzkiego, pp. 145–68.

Norman, Alex, and Johnson Mark. 2011. World Youth Day: The creation of a modern pilgrimage event for evangelical intent. *Journal of Contemporary Religion* 26: 371–85. [CrossRef]

Nowotny, Sławomir. 2016. *Odbiór Światowych Dni Młodzieży i Jego Kulturowe Konteksty [Perception of World Youth Day and Its Cultural Contexts]*. Warszawa: Narodowe Centrum Kultury & Instytut Statystyki Kościoła Katolickiego.

Paradise in the City. 2019. Available online: https://paradiseinthecity.com (accessed on 14 September 2019).

Pfadenhauer, Michaela. 2010. The eventization of faith as a marketing strategy: World Youth Day as an innovative response of the Catholic Church to pluralization. *International Journal of Nonprofit and Voluntary Sector Marketing* 15: 382–94. [CrossRef]

Reader, Ian. 2015. *Pilgrimage: A Very Short Introduction*. Oxford: Oxford University Press.

Richards, Greg, and Palmer Robert. 2010. *Eventful Cities: Cultural Management and Urban Revitalization*. Oxford: Butterworth-Heinemann.

Ritchie, J. R. Brent, and Beliveau Donald. 1974. Hallmark events: An evaluation of a strategic response to seasonality in the travel market. *Journal of Travel Research* 14: 14–20. [CrossRef]

Ritchie, J. R. Brent. 1984. Assessing the impacts of hallmark events: Conceptual and research issues. *Journal of Travel Research* 23: 2–11. [CrossRef]

Rymarz, Richard. 2008. Its Not Just Me Out There: Type A Pilgrims at World Youth Day. *Australian e-Journal of Theology* 12: 1–18.

Sarniewicz, Karolina, and Salawa Anna. 2016. *Zróbmy Raban! Niezbędnik na Światowe Dni Młodzieży [Let's Make a Rumpus! A Must-Have for the World Youth Day]*. Kraków: Znak.

Seweryn, Renata. 2018. Motivations of volunteers and their role in the tourist promotion of the location of World Youth Day (on the example of Krakow). *Ekonomiczne Problemy Turystyki* 1: 139–47. [CrossRef]

Singleton, Andrew. 2011. The impact of World Youth Day on religious practice. *Journal of Beliefs & Values* 32: 57–68.

Uysal, Muzaffer, Gahan Lawrence, and Martin Bonnie. 1993. An examination of event motivations: A case study. *Festival Managment and Event Tourism* 1: 5–10.

Webber, Ruth. 2008. The social milieu of World Youth Day 08 and Pilgrims' social ethics'. Paper presented at Annual Meeting of the Society for the Scientific Study of Religion, Louisville, KY, USA, October 16–19.

Zduniak, Agnieszka. 2010. Event jako ponowoczesna forma uczestnictwa w życiu społecznym [Event as a postmodern form of participation in social life]. *Roczniki Nauk Społecznych* 2: 207–34.

Zduniak, Agnieszka. 2015. Przyczyny popularności religijnych eventów w perspektywie teorii potrzeb ludzkich Abrahama Maslowa [Reasons for the popularity of religious events in the perspective of the human needs theory of Abraham Maslow]. *Colloquium Wydziału Nauk Humanistycznych i Społecznych* 3: 251–67.

Article

Religious Festival Marketing: Distinguishing between Devout Believers and Tourists

Kuo-Yan Wang [1], Azilah Kasim [2,*] and Jing Yu [1]

[1] Department of Marketing in School of Economics and Management, Guangdong University of Petrochemical Technology, Maoming 525000, China; kywang@gdupt.edu.cn (K.-Y.W.); yujing@gdupt.edu.cn (J.Y.)

[2] School of Tourism, Hospitality & Event Management, Universiti Utara Malaysia, Sintok 06010, Malaysia

* Correspondence: azilah@uum.edu.my

Received: 2 July 2020; Accepted: 10 August 2020; Published: 12 August 2020

Abstract: Customer classification is an integral part of marketing planning activities. Researchers have struggled to classify "pilgrims" and "tourists" because these groups overlap to a large extent in terms of their identities while participating in religious activities/sightseeing. To achieve sustainable tourism development for the region with rich religious and cultural characteristics, the present article outlines a process for analyzing the motivation of participants attending religious festival of Mazu in Taiwan and then classifies religious festival participants according to their motivations. Using cluster sampling, a total of 280 responses were obtained and analyzed. The results revealed four different motivation categories: Fun traveler, devout believer, cultural enthusiast, and religious pragmatist. The study concludes that while festivalgoers are influenced by secularization to some extent, the original doctrine of the religion epitomized in the festivals fundamentally retains the essence and spirit of its religious rituals. The findings may have a significant value for the development of religious tourism marketing as it offers a foundation for future research seeking to develop regional cultural and religious sightseeing attractions sustainably.

Keywords: festival; customer classification; factor analysis; motivation; folklore belief

---

1. Introduction

To stimulate sustainable brand development, firms usually exert great efforts to classify their customers, allowing them to identify the most profitable target markets and optimally target greater numbers of such customers. Customer classification involves the subdivision of a market into discrete customer groups that share similar characteristics, which can be a powerful means of identifying unmet customer needs (Onwezen 2018). Companies that identify underserved segments can subsequently outperform the competition by developing unique products and services. Generally, customer classification can help a business develop marketing campaigns and pricing strategies to extract maximum value from both high- and low-profit customers (Bhatnagar and Gopalaswamy 2017). A company, as well as an ad-hoc organization, can use customer classification as the principal basis for allocating resources to product development, marketing, service, and delivery programs (Timoshenko and Hauser 2019; Wang et al. 2018).

Organizations increasingly use festivals as a means to promote tourism and stimulate the regional economy due to the added value that festivals offer: The potential to reinforce local economies, encourage the conservation of festival-themed commodities, sustain and preserve local identities, and create opportunities for tourists to engage with local cultures and people (Viljoen et al. 2017; Berg and Sevón 2014; Saayman and Saayman 2006). For festival-related organizations, therefore, an accurate customer classification can help marketers better target customers during the planning phases, thus attracting more sponsors and income.

Festivals are a type of cultural event, highlighting the unique features of a destination or theme (Zhang et al. 2019). The explosion in festival numbers has multifaceted causes. Supply factors have an influence, including cultural-based event planning, tourism industry development, and civic movement repositioning (Bakas et al. 2019; Duxbury et al. 2019). Demand factors also play a role, including serious leisure, lifestyle sampling, socialization needs, and the desire for creative and firsthand experiences (Lyu and Lee 2016). Festivals can have many themes, such as music (Li and Wood 2016; Rowley and Williams 2008), sightseeing attractions (Getz and Page 2016; Getz 2010; Lee et al. 2004), and sport competitions (Brown et al. 2016; Brown 2007). Previous research has discussed, in-depth, visitors' motivations for attending festivals and special events (Ruiz et al. 2019; Grappi and Montanari 2011; Kim et al. 2006; Prentice and Andersen 2003; Dewar et al. 2001; Crompton and McKay 1997). However, few studies specifically address religious festivals, as it is difficult to determine participants' motivations for attending such events. Thus, the application of appropriate strategies for festival marketing programs is a challenge for activity sponsors, festival organizers, and tourism sector planners. In these cases, the visitor is viewed as being just along for the ride in a defocused carnival environment.

The phrase *matsuri* in the Japanese language denotes "serving" and "entertaining" the deity in Japanese religion (Kitagawa 1988). The appearance or presentation of *matsuri*, a ritual or ceremony that serves as the embodiment of meaning, effectively establishes morality, and religious festivals around the globe can last for long periods. In Kawano's (2005) definition of ritual practice, rituals are traditionally presented to ordinary people seeking to gain something from their voluntary participation. In practical terms, religious rituals themselves constitute a central force of community cohesion as they can attract wide community attention and participation. One such ritual is the annual festivals of the Hitachinokuni Sosha Shrine in Ishioka City, Ibaraki Prefecture, and Giōn Matsuri in Kyoto Prefecture, Japan (Kim 2005). Similarly, folkloric beliefs, rituals, and ceremonies in Chinese tradition express deity reverence and have been practiced for several hundred years in Taiwan. From the perspective of a social-functional interpretation, an efficient religious festival also helps regional religions and cultural developments. Shimazono (1998) asserted that the idea of the general acceptance of a religion stems from religious organizations and their followers, who claim that participation in religious activities is voluntary. Participants tend to follow, to a certain extent, suggestions regarding what constitutes a suitable donation. Nevertheless, the promotional efforts of religious organizations, such as the sale of objects of worship, incense sticks, amulets, and deity-themed objects, are frequently visible. Since people widely believe that the purchase of these objects guarantees an end to misfortune, the danger of dishonest behavior in the pursuit of personal profit exists (Wang 2019; Wang 2014a). The marketing of religious commodities (e.g., religious objects, merchandise, or souvenirs) may also elicit similar attitudes where, while an underlying sacred meaning remains, there is no objection towards the commercialization of religious products (Zaidman 2003). The literature, however, does not specifically discuss this in the context of religious events such as religious cultural festivals or deity birthdate ceremonies. In addition, there is still very little discourse on the motivations of the participants of such an event. Therefore, this empirical study focuses on this gap of knowledge using a Chinese Taoist temple festival organizer in Taiwan as its research setting.

The rest of this paper is organized as follows. We briefly present the most influential Tao deity, Mazu, worshiped in Taiwan. This figure remains an indispensable part of Chinese cultural life because Taiwanese folkloric beliefs are heavily influenced by the Mainland Chinese provinces of Fujian and Guangdong, as exemplified by the ancestral homes built during the immigration wave of the 17th through the 19th centuries (Wang 2014a). For an improved customer classification of sustainable regional religious or culture-themed tourism development, the next section describes the importance of the motivation of participants in religious activities. In addition, the results of a factor analysis, in terms of basic statistics for participants' motivation in religious festivals, are presented. The final section concludes this empirical study.

2. Research Focuses

2.1. Mazu Belief Practices in Taiwan

Diachronic folkloric beliefs in Chinese society and religious culture have developed over the past few thousand years and are a mixture of Buddhism, Confucianism, and Taoism. Belonging to the same cultural circle and tradition of philosophical thought, religions in Taiwan are syncretized and pantheistic (Wang 2014a; Chang 2009). In addition to the inherent aboriginal culture (Liao et al. 2019), Taiwanese folkloric beliefs are heavily influenced by Fujian and Guangdong, two provinces in Mainland China, because of the ancestral homes built during the immigration waves between the 17th and 19th centuries. As the immigrant population adjusted to a new living environment and an uncertain future, communal rituals, and folkloric beliefs derived from the mainland immigrants' original religions became widespread on the island (Wang 2014a). These festivals have shaped the belief system of contemporary ethnic Chinese residents now living in Taiwan. More than 10,000 Taoist, Buddhist, and Confucian religious sites exhibit the dominant Chinese pantheistic belief structure.

Folkloric beliefs are a form of religion that recognizes all deities as gods. Temples dedicated to the deities commonly form alliances with several temples of similar characteristics dedicated to the same deities. According to Chinese pantheistic beliefs, the deity represents the soul or spirit. This belief may derive from Taoist religious communal rituals. As the most influential goddess of the sea in China, Mazu is at the center of beliefs and customs, including rituals, ceremonies, and folk practices, throughout coastal areas, with millions of followers (Chang et al. 2012). Scholars have argued that the role of Mazu is more than that of a guardian angel for fishermen (Guo et al. 2006; Shuo et al. 2009). The Mazu belief, as a social cohesion catalyst in Chinese society, was inscribed on the representative list of the Intangible Cultural Heritage of Humanity, of the United Nations Educational, Scientific, and Culture Organization (UNESCO) in 2009 (UNESCO and The Mazu Belief and Customs 2009). Pilgrims gather in the so-called holy land for specific festivals, such as that held on March 23 of every lunar year, to celebrate the birthday of Mazu. Many related activities accompany this commemoration, such as group worship ceremonies and pilgrimages to locations where the temple is enshrined. Even during the period of Japanese colonial occupation, the rulers did not challenge the power of Mazu beliefs. The rulers of the Taiwan Colonial Government showed appreciation for Mazu's mercy, with the additional purpose of winning over different ethnic groups. Currently, for politicians—from island wide leaders to village headmen, regardless of their own beliefs—visiting and praying in famous temples represents a way to win the votes, hearts, and minds of the people. After years of development, Mazu now acts as a bridge for numerous politicians joining Mazu-themed festivals. From tourism point of view, Mazu celebrations are important tourism income earners as they could attract millions of visitors willing to spend nine days marching through different cities just to celebrate the deity's birthday (Wong 2013, April 12).

Pilgrimage enables tourism-related activities and is an important regional characteristic because it embodies the strong local folk culture. In the case of Dajia Mazu Pilgrimage, the festival involves a nine-day parade carrying Mazu statue in a chair for 12 h a day through mountains and rough terrains. The parade starts and ends at the Jenn Lann Temple in the city but will pass through more than 100 temples at the coastal counties of Changhua, Yunlin and Chiayi in its 300 km journey (Wong 2013, April 12). It is said that every temple stop provides interaction opportunities between Mazu Goddess with the Mazu statue that resides in that particular temple, that empowers the temple and brings luck to the local community. Approximately 200,000 pilgrims are involved in each Dajia Mazu Pilgrimage. They receive free food and sleep everywhere (the local community's gardens, garages, living rooms etc.) for free at each of the stops (Tsao 2014).

Though the festival was the result of public–private cooperation in Taichung city (Tsao 2014), the steady increase of participants has led to a booming growth of tourism, cultural activities, artistic activities as well as local food and transportation businesses throughout the event. In 2013, the total number of participants was five millions people (Wong 2013, April 12). This essentially highlights the tourism importance of the event which brings with it the issue of commodification of religious tourism, where mythical narration is combined with non-religious demands, providing tourism opportunities for areas surrounding pilgrimage sites (Bixby 2006). In recent years, many temples in Taiwan have strongly promoted folkloric beliefs in cultural development to fascinate believers and tourists, thereby increasing donations and marketing revenues, including the creative product development of religious items. Figure 1 displays deity-themed cultural creative goods (e.g., amulets, postcards, key rings, and USB drives) in vending machines at a historical temple in Southern Taiwan.

Figure 1. A vending machine with religious-themed creative goods in a Taiwanese temple (Photo from the author).

2.2. *Classifying Religious Festival Participants: Filling in the Blanks*

In general, non-profit organizations have more of a need for marketing than for-profit organizations because the former mainly derive their revenues from social sponsorships (Bowdin et al. 2006). In addition, a favorable public image facilitates capital mobility (Moore 2000). Chinese temples have a special classification that falls somewhere between for-profit and non-profit, as they mainly receive funding from worshiper contributions, with some subsidies from local authorities. A temple needs to allocate resources to gain social recognition and attract more loyal believers (customers) to survive in a fiercely competitive and highly overlapping environment (Iriobe and Abiola-Oke 2019; Collins-Kreiner 2018; Wang 2014b). In recent years, temples have begun to carry out an increasing number of regional religious and culture-based activities, which attract more visitors, lead to the development of the local tourism sector and peripheral industries and strengthen the temple's reputation (Wang 2014b). Therefore, optimizing services to attract believers/customers/donators has become a top priority (Wang 2014a). That being said, as Prentice and Andersen (2003) noted, "Not everyone at a destination during a festival can be assumed to be a festival-goer." It is difficult to correctly classify pilgrims and tourists. A further question is whether or not the commercial development of religious festivals has harmed the spirituality inherent in the associated religious rituals and belief structures.

Scholars have discussed the differences and similarities between tourists and pilgrims (Durán-Sánchez et al. 2018; Mwebaza et al. 2018; Chang et al. 2012). Both tourists and pilgrims require discretionary income, leisure time, and social sanctions permissive of travel (Durán-Sánchez et al. 2018). Secularization of religious communication and development is inevitable in the modern era (Kluver and Cheong 2007) thereby making it increasingly difficult to distinguish between pilgrims and tourists. Accordingly, facing the modern secularization, the behaviors, rituals, and motivations of religious festival participants may have been adjusted. Cohen (1992) for example, has argued that visitors to a pilgrimage site may actually not be followers of that site's religion and therefore should be classified as just ordinary travelers.

The term pilgrimage itself comes to denote a leisure activity without a firm religious purpose (Liutikas 2017; Matoga and Pawłowska 2018). Pilgrimages have powerful political, economic, social, and cultural implications and can even affect global trade and social phenomena. Pilgrimage inevitably necessitates spatial movement, and such a form of "circulation" has no less an effect on the environment (Collins-Kreiner 2010; Kim et al. 2016; Moufahim and Lichrou 2019). Moreover, the concept of a "large-scale acquaintance network" distributed through the power of the Internet (Tomochi 2010; Tanaka 2014) is key. The acceleration of information exchange and social networks in small-scale societies occurs easily, and large-scale informal organizations cannot be ignored (Balbo et al. 2016) because many religious communities have utilized the Internet as part of their religious mission and growth strategy (Kluver and Cheong 2007).

The above-mentioned trend may have encouraged many people to focus attention on their own regional cultural and religious heritage preservation activities (Chen 2016). This trend could explain people's motivations to participate in religious festival activities and simultaneously highlight the importance of effective pilgrim/tourist classification. However, an in-depth discussion of actual, more complex phenomena related to distinguishing the real motivations behind a pilgrim or a tourist attending a religious activity is still lacking. Hence this study attempts to address the question—when a religious festival such as the Mazu festival becomes more commercial-driven, does it affect participants' motivations as well? The question of whether or not intensive marketing and promotion can affect 'religious purity' or the original reason for participating in a religious event (Shiomura 2009) is worth discussing. That is why this study aims to address the long-standing problem of distinguishing between pilgrims and tourists in religious festivals in attempt to provide a more comprehensive view of the issue and its implications for temple managers as organizers.

3. Research Methods

3.1. Instruments

The survey instrument was based on that used by Crompton and McKay (1997), which is considered to be an effective analysis measurement tool. This instrument, which is based on segmenting visitors and identifying their specific needs and behaviors, enables festival organizers to develop effective marketing strategies and, ultimately, ensure their long-term economic viability. Therefore, numerous studies on motivations within festival research have used this tool to investigate World Heritage Sites (Woosnam et al. 2018), food-themed events (Özdemir and Seyitoğlu 2017), sexual tourism (Ying and Wen 2019), special holidays (Mariani and Giorgio 2017), and so on.

The instrument starts with information on respondents, such as festival visit times, religious beliefs, and where respondents heard about the event, followed by the motive-related items for observing respondents' attitudes toward religious festival participation. The responses were analyzed to make an appropriate classification using factor analysis.

Numerous articles regarding marketing research have applied multivariate methods to problems of customer classification. Factor analysis techniques are among the many potentially useful methods for researchers (Takele 2019). Many academics have recognized factor analysis as a simple-to-use survey method with a great deal of practical value, particularly for market segmentation and customer classification research (Boivin and Tanguay 2019; Lee and Kwag 2019; Dhakal et al. 2017).

For this empirical study, we applied factor analysis techniques to classify participants of several Mazu temple festivals held in Taiwan, thereby demonstrating religious festival visitor typology in a realistic setting. The advantages of using quantitative research are that it is suitable for larger sample sizes, has high objectivity and accuracy, saves time in information collection and processing, and is more cost-effective. Hence, the results were obtained by the research assistants hired for this study, who issued and collected questionnaires as well as conducted the preliminary data analysis.

These classifications were made based on 12 questionnaire items. This form of factor analysis first extracts mutually independent factors and then classifies the respondents' motivations based on their scores on these factors. The results, obtained through factor analysis, can help us better understand the phenomena and implications of interest, and in doing so, one must always place a study within the context of existing theory and practical applications.

3.2. Research Subjects and Data Collection

A total of five temples were used in March 2019 to understand Mazu-themed festival performances across the island. The temples are located in the northern part of Taiwan (temple A), in the central region (temples B and C), and the south (temples D and E). Taking into consideration tax issues and other economic factors, four temples refused to disclose their identity but accepted to take part in the empirical survey. Each temple was accustomed to annually organizing Mazu-themed festivals. Similar to the mass participation ceremony on the deity's important day, on a particular date, the deity makes a procession to affiliated temples several hundred kilometers away, accompanied by a mass of volunteer followers seeking blessings. The most famous example of this practice is for Mazu, dedicated by the Jenn Lann Temple in Dajia, Taichung City. A grand procession to the affiliated Mazu temple, Feng Tian Temple in Singang, Chiayi County, occurs each March of the lunar year with numerous devout believers involved, which also adds to the prosperity of the local economy (Wang 2014a).

In practical terms, religious object commercialization is already a worldwide trend, especially for the enhancement of regional tourism sectors. Therefore, in addition to the preliminary classification of festival visitors as pilgrims and tourists, we also investigated whether the phenomenon of religious object commercialization harms the spirituality of individuals. To seek the answer to the questions, we also surveyed temple B's Mazu-themed festival participants to understand the motivation of the Mazu beliefs of pilgrims and tourists. The survey period was from 23 March to 6 April 2019, during an activity involving approximately 30,000 participants. Cluster sampling was applied to randomly chosen tourists/pilgrims of temples B and C; one researcher conducted the survey at each temple. A total of 330 questionnaires were issued and filled out in person. The survey ruled out respondents who were unwilling to answer and invalid responses, leaving 280 valid interviewees (n = 280, approximately 84.8%) in this study. Of the sample, as illustrated in Table 1, 62.5% (175) were female, and 37.5% (105) respondents were male. A total of 40.35% (113) were 50 years of age or above, 35% (98) were 30–39 years old, and 20% (56) were under the age of 29. Regarding the number of visits to the deity-themed festival, most respondents (212) had participated more than two times (75.8%). Source of information about the festival for 76.1% (213) of the interviewees was the Internet, social media (e.g., Facebook, Twitter, Line, and WeChat); 17.1% learned about the event from TV commercials (48). Of the sample, 71.4% (200) pointed out they were primarily of a folkloric belief, which is a mixture of Buddhism, Confucianism, and Taoism, a widespread traditional faith in the Chinese community.

Table 1. Profile of respondents (n = 280).

Variables	Sample Size	Percentage (%)	Variables	Sample Size	Percentage (%)
Gender			**Religion**		
Male	105	37.5	Buddhism (monk or lay)	53	18.9
Female	175	62.5	Christianity	3	1.1
			Folklore belief/Taoism	200	71.4
Age			Islam	1	0.4
19 and below	13	4.6	No specific	23	8.2
20–29	43	15.4			
30–39	98	35.0	**Number of past participations in Mazu festival**		
40–49	63	22.5	Once and never	68	24.2
50–59	43	15.4	Twice	45	16.1
60 and above	20	7.1	Three times	120	42.9
			Four times or more	47	16.8
Source of festival information					
Social media	213	76.1	**Numbers of travel mate**		
TV commercials	48	17.1	Family member	138	49.3
Family member or friend	12	4.3	Friend	128	45.7
Festival organizer	7	2.5	None	14	5.0

4. Four Types of Religious Festival Participants

To effectively classify respondents' motivations to attend a religious event, the researcher also employed a 5-point Likert scale (1 = *strongly disagree* to 5 = *strongly agree*). Factor analysis was used to identify the constructs underlying the 12 posited festival participant motivations from modified questionnaire items originally designed by Crompton and McKay (1997). The questionnaire construction fully considered internal consistencies for each survey item, as well as reliability and validity less than the normal value (Cronbach's alpha value = 0.72). Each item reflected the list of selected topics and was significantly different at 5%. The KMO value was given as 0.83, meaning that each item could be validly analyzed (Hemmati et al. 2018; Akhoondnejad 2016). Table 2 illustrates the factors influencing the participants of Mazu-themed activities, and the corresponding motivations can be divided into four main types according to the festival participants' motivations: leisure travelers, devout believers, culture enthusiasts, and religious pragmatists.

Table 2. The motivations of religious activity participant—factor analysis result.

Self-Description	Leisure Traveler	Devout Believer	Cultural Enthusiast	Religious Pragmatist
I was attracted by the large amount of activity advertisements	**0.819**	0.149	0.259	0.083
This activity is like a carnival, I like this atmosphere	**0.723**	0.099	0.343	0.046
I am coming here for this activity, at the same time, I can visit my friends and relatives	**0.849**	0.143	0.114	0.092
I am travelling here	**0.771**	0.043	0.035	0.134
I am here to salute and worship to Mazu	0.093	**0.839**	0.128	0.049
My family members are all Mazu followers so I participate in this activity	0.086	**0.808**	0.178	0.184
Because it is the inherent custom in this place	0.218	0.068	**0.696**	0.038
The temple is filled with a sense of history	0.252	0.093	**0.641**	0.102
I believe Mazu will make me and my family happy and joy if I join the ceremony	0.221	0.292	0.282	**0.537**
The temple's ancillary products (e.g., deity dolls, clothes, hats, amulets) are distinctive	0.252	0.041	0.388	**0.613**
I always attend this activity; it is so much fun	0.167	0.169	0.204	0.360
I am coming for fulfill the promise of Mazu blessings	0.211	0.033	0.386	0.113
Total respondents (n = 280)	70	68	70	72

As seen in Table 2, the factor analysis results revealed that each factor number over 0.5 indicates that the interviewees widely recognized the item. The number of respondents for each type was almost equal. Hence, we classified four basic types of motivation as follows: Leisure travelers, devout believers, cultural enthusiasts, and religious pragmatists.

The leisure traveler accounted for most of the factors. The purpose of such participants consists of having fun travel experiences. They participate in religious activities for leisure and entertainment. Leisure travelers prefer a lively atmosphere, such as the carnival celebrations in Rio de Janeiro, Brazil. In other words, the term *tourist* for a leisure traveler is much more common than the term *pilgrim*. Conversely, other participants and family members strongly put their faith in the Mazu belief. Devout believers are active participants attracted by the temple's external promotions. They are undoubtedly the typical pilgrims.

The original intentions of the remaining two types of participants, cultural enthusiasts and religious pragmatists, are difficult to classify as "tourist" or "pilgrim" intentions. Cultural enthusiasts are commonly historian types, passionate about gaining more historical and cultural knowledge; compared to other types of participants, cultural enthusiasts are more interested in the origin and knowledge of local historical religions and customs. In Taiwan, famous temples play a central role in local folkloric beliefs. Therefore, they are also full of history and cultural heritage that motivates the participation of cultural enthusiasts. The starting point of cultural enthusiasts, however, is not necessarily on a religious level. In addition, despite the common perception of temple visitors being "belief followers," religious pragmatists tend to recognize that participation in religious activities benefits them in terms of attending Mazu ceremonies and buying associated products (e.g., dolls, clothes, hats, and amulets) as tools to help them realize their wishes. Apparently, their motivation is to meet personal spiritual and material conditions. Furthermore, commercial advertisements influence leisure travelers and religious pragmatists to a higher degree. The survey results indicate that participants were significantly attracted by social network and TV advertising from deity-themed festivals held by a Taoist temple in Taiwan (93.2% of respondents). Moreover, this article successfully classified the four types of religious festival participants between the tourist and pilgrim dimensions in Taiwan. The classification of participants' motivations also explains the current Mazu-themed activity in the festival and takes into account tourism activities related to religious beliefs and cultural characteristics, thus attracting participants with diverse motivations.

5. Implications

We successfully classified the four basic typologies of the motivations of Mazu festival participants and strengthened the marketing classification perspective in the case of annual temple festival participants. Several key points of this research are discussed below.

For the case of leisure-oriented visitors, Swatos (2006) provided an unprecedented perspective on "pilgrimage for eminent persons," and Norman (2011) extended this idea to explore phenomena such as sincere worship and praying to God. In other words, no matter "whom" (the object receiving worship), the object only needs to meet visitors' spiritual needs (e.g., family affections, friendship, love, or happiness) and can be "created" or "shaped." Similarly, the event may also be invented or emphasize historical legendary narratives.

Next, many visitors learned about the Mazu festival from social media and TV promotions. Regarding doubts about "entertaining" religious activities, such as those promoted by for-profit organizations, many temples are seemingly profit oriented due to the high level of competition in Taiwan, which also urges them to actively promote themselves via TV advertising, online communities, ritual narratives, and even ad placements and marketing. Tourism with a cultural theme has become a good entry point for the temple studied in this research. This approach also helps the temple meet the leisure or spiritual needs of worshipers or travelers, as well as to stimulate local economic and tourism development to achieve a mutually beneficial effect.

Third, our factor analysis addresses four types of religious festival participants according to the passage relating to the respondents' motivations. Some travelers who joined the group ceremony where they were motivated by a desire to participate in a public ceremony or activity. Conversely, some visitors were driven more by religious pursuits than by their egoistic needs. In this study, the four types of participants coexist in the religious event. Regarding the differentiation between tourists and pilgrims, despite the small survey sample, the researcher managed to make a preliminary classification of those participants who could implement more precisely targeted sales for future festival marketing events.

6. Conclusions

In this study, the researcher performed an exploratory case study of folklore belief temples in Taiwan by studying attendants at the Mazu-themed festival in April 2019. The survey uncovered respondents' motivations for participation. These motivations were classified into four types based on the classification developed by Crompton and McKay (1997): Leisure travelers, cultural enthusiasts, religious pragmatists, and devout believers. This classification can serve as a valuable reference for subsequent studies.

Communities can open themselves to the phenomenon of secularization and, at the same time, maintain the original intention of the religious doctrine. Furthermore, the survey results show that the influence of the Internet was strongly significant for the two sampled festival marketing activities in 2019. Such marketing strategies are particularly attractive for the young generation and may help address some issues to meet the needs of religious festival participants. Additionally, this empirical study features a quantitative method because of time and cost limitations. Qualitative analysis can also be considered in future research to draw more comprehensive results and conclusions.

The cross-sectional data used for this study may have limited the findings in the sense that if the survey were administered at other times or other locations, the findings may be different. Hence it is proposed future researchers try to avoid this by engaging in a more comprehensive or longitudinal survey. We also suggest that future researchers should examine and segment different religions and locations to extend this preliminary classification, particularly new religious groups founded after World War II, for participants of various activities (e.g., worship, spiritual practices, and festivals). Further studies on other locations are needed, and much work is still required to analyze and empirically test these new ideas.

Author Contributions: All authors contributed equally to this work. All authors wrote, reviewed, and commented on the manuscript. All authors have read and approved the final manuscript.

Funding: This research was funded by Talent Introduction Project of Guangdong University of Petrochemical Technology, "Research on Marketing Planning and Practice of Cultural Theme Tourism Industry Circle".

Conflicts of Interest: The authors declare no conflict of interest.

References

Akhoondnejad, Arman. 2016. Tourist loyalty to a local cultural event: The case of Turkmen handicrafts festival. *Tourism Management* 52: 468–477. [CrossRef]

Bakas, Fiona Eva, Nancy Duxbury, and Tiago Vinagre de Castro. 2019. Creative tourism: Catalyzing artisan entrepreneur networks in rural Portugal. *International Journal of Entrepreneurial Behavior & Research* 25: 731–52. [CrossRef]

Balbo, Andrea L., Erik Gómez-Baggethun, Matthieu Salpeteur, Arnald Puy, Stefano Biagetti, and Jürgen Scheffran. 2016. Resilience of small-scale societies: A view from drylands. *Ecology and Society* 21: 53. [CrossRef]

Berg, Per Olof, and Guje Sevón. 2014. Food-branding places–A sensory perspective. *Place Branding and Public Diplomacy* 10: 289–304. [CrossRef]

Bhatnagar, Navneet, and Arun Kumar Gopalaswamy. 2017. The role of a firm's innovation competence on customer adoption of service innovation. *Management Research Review* 40: 378–409. [CrossRef]

Bixby, Brian. 2006. Consuming simple gifts: Shakers, visitors, goods. In *The Business of Tourism: Place, Faith, and History*. Edited by Philip Scranton and Janet F. Davidson. Philadelphia: University of Pennsylvania Press, pp. 85–108, 259–64.

Boivin, Maryse, and Georges A. Tanguay. 2019. Analysis of the determinants of urban tourism attractiveness: The case of Québec City and Bordeaux. *Journal of Destination Marketing & Management* 11: 67–79.

Bowdin, Glenn, Johnny Allen, William O'Toole, Rob Harris, and Ian McDonnell. 2006. *Events Management*, 2nd ed. Oxford: Butterworth-Heinemann.

Brown, Graham. 2007. Sponsor hospitality at the Olympic Games: An analysis of the implications for tourism. *International Journal of Tourism Research* 9: 315–27. [CrossRef]

Brown, Graham, Andrew Smith, and Guy Assaker. 2016. Revisiting the host city: An empirical examination of sport involvement, place attachment, event satisfaction and spectator intentions at the London Olympics. *Tourism Management* 55: 160–72. [CrossRef]

Chang, Wen-Chun. 2009. Religious attendance and subjective well-being in an Eastern-culture country: Empirical evidence from Taiwan. *Marburg Journal of Religion* 14: 1–30.

Chang, Horng-Jinh, Kuo-Yan Wang, and Shean-Yuh Lin. 2012. When backpacker meets religious pilgrim house: Interpretation of oriental folk belief. *Journal for the Study of Religions and Ideologies* 11: 76–92. Available online: http://www.jsri.ro/ojs/index.php/jsri/article/viewFile/614/545 (accessed on 2 January 2020).

Chen, Wei-Chih. 2016. Leisure participation, job stress, and life satisfaction: Moderation analysis of two models. *Social Behavior and Personality: An International Journal* 44: 579–88. [CrossRef]

Cohen, Erik. 1992. Pilgrimage and tourism: Convergence and divergence. In *Sacred Journeys: The Anthropology of Pilgrimage*. Edited by Alan Morinis. Westport: Greenwood Press, pp. 47–61.

Collins-Kreiner, Noga. 2010. The geography of pilgrimage and tourism: Transformations and implications for applied geography. *Applied Geography* 30: 153–64. [CrossRef]

Collins-Kreiner, Noga. 2018. Pilgrimage-tourism: Common themes in different religions. *International Journal of Religious Tourism and Pilgrimage* 6: 8–17.

Crompton, John. L., and Stacey L. McKay. 1997. Motives of visitors attending festival events. *Annals of Tourism Research* 24: 425–39. [CrossRef]

Dewar, Keith, Denny Meyer, and Wen Mei Li. 2001. Harbin, lanterns of ice, sculptures of snow. *Tourism Management* 22: 523–32. [CrossRef]

Dhakal, Basanta, Azaya Bikram Sthapit, and Shankar Prasad Khanal. 2017. Factor analysis of local residents' perceptions towards social impact of tourism in Nepal. *International Journal of Statistics and Applied Mathematics* 2: 1–8.

Durán-Sánchez, Amador, José Álvarez-García, Maria De la Cruz Río-Rama, and Cristiana Oliveira. 2018. Religious tourism and pilgrimage: Bibliometric overview. *Religions* 9: 249. [CrossRef]

Duxbury, Nancy, Fiona Eva Bakas, and Cláudia Pato de Carvalho. 2019. Why is research–practice collaboration so challenging to achieve? A creative tourism experiment. *Tourism Geographies*, 1–26. [CrossRef]

Getz, Donald. 2010. The nature and scope of festival studies. *International Journal of Event Management Research* 5: 1–47.

Getz, Donald, and Stephen J. Page. 2016. Progress and prospects for event tourism research. *Tourism Management* 52: 593–631. [CrossRef]

Grappi, Silvia, and Fabrizio Montanari. 2011. The role of social identification and hedonism in affecting tourist re-patronizing behaviours: The case of an Italian festival. *Tourism Management* 32: 1128–40. [CrossRef]

Guo, Yingzhi, Samuel Seongseop Kim, Dallen J. Timothy, and Kuo-Ching Wang. 2006. Tourism and reconciliation between Mainland China and Taiwan. *Tourism Management* 27: 997–1005. [CrossRef]

Hemmati, Farhad, Fatemeh Dabbaghi, and Ghahraman Mahmoudi. 2018. Iran's need for medical tourism development. *Journal of Research in Medical and Dental Science* 6: 269–73.

Iriobe, Ofunre, and Elizabeth Abiola-Oke. 2019. Moderating effect of the use of eWOM on subjective norms, behavioural control and religious tourist revisit intention. *International Journal of Religious Tourism and Pilgrimage* 7: 38–47. [CrossRef]

Kawano, Satsuki. 2005. *Ritual Practice in Modern Japan: Ordering Place, People, and Action*. Honolulu: University of Hawai'i Press.

Kim, Hyunjung. 2005. Structure and significance of the Japanese urban festival: Diachronic and synchronic analyses of structure of the annual festival of Hitachinokuni Sosha shrine in Ishioka City. *Journal of Korean Society for Cultural Anthropology* 38: 45–81.

Kim, Hyounggon, Marcos C. Borges, and Jinhyong Chon. 2006. Impacts of environmental values on tourism motivation: The case of FICA, Brazil. *Tourism Management* 27: 957–67. [CrossRef]

Kim, Bona, Seongseop Sam Kim, and Brian King. 2016. The sacred and the profane: Identifying pilgrim traveler value orientations using means-end theory. *Tourism Management* 56: 142–55. [CrossRef]

Kitagawa, Joseph M. 1988. Some Remarks on Shintō. *History of Religions* 27: 227–45. [CrossRef]

Kluver, Randolph, and Pauline Hope Cheong. 2007. Technological Modernization, the Internet, and Religion in Singapore. *Journal of Computer-Mediated Communication* 12: 1122–42. [CrossRef]

Lee, Jung Wan, and Michael Kwag. 2019. Structural equation modelling with second-order confirmatory factor analysis: Critical factors influencing consumer behavior in medical tourism. In *Quantitative Tourism Research in Asia*. Singapore: Springer, pp. 223–43.

Lee, Choong-Ki, Yong-Ki Lee, and Bruce E. Wicks. 2004. Segmentation of festival motivation by nationality and satisfaction. *Tourism Management* 25: 61–70. [CrossRef]

Li, Yan-Ning, and Emma H. Wood. 2016. Music festival motivation in China: Free the mind. *Leisure Studies* 35: 332–51. [CrossRef]

Liao, Chu-Chu, Ying-Xing Lin, and Huey-Hong Hsieh. 2019. Satisfaction of indigenous tourism from residents' perspective: A case study in Nantou county, Taiwan. *Sustainability* 11: 276. [CrossRef]

Liutikas, Darius. 2017. The manifestation of values and identity in travelling: The social engagement of pilgrimage. *Tourism Management Perspectives* 24: 217–24. [CrossRef]

Lyu, Seong Ok, and Hoon Lee. 2016. Latent demand for recreation participation and leisure constraints negotiation process: Evidence from Korean people with disabilities. *Journal of Leisure Research* 48: 431–49. [CrossRef]

Mariani, Marcello M., and Luisa Giorgio. 2017. The "Pink Night" festival revisited: Meta-events and the role of destination partnerships in staging event tourism. *Annals of Tourism Research* 62: 89–109. [CrossRef]

Matoga, Łukasz, and Aneta Pawłowska. 2018. Off-the-beaten-track tourism: A new trend in the tourism development in historical European cities. A case study of the city of Krakow, Poland. *Current Issues in Tourism* 21: 1644–69. [CrossRef]

Moore, Mark. H. 2000. Managing for value: Organizational strategy in for-profit, nonprofit, and governmental organizations. *Nonprofit and Voluntary Sector Quarterly* 29: 183–204. [CrossRef]

Moufahim, Mona, and Maria Lichrou. 2019. Pilgrimage, consumption and rituals: Spiritual authenticity in a Shia Muslim pilgrimage. *Tourism Management* 70: 322–32. [CrossRef]

Mwebaza, Susan, Julius Jjuuko, and Provia Kese. 2018. Religious Tourism and Pilgrimage: Conflicts and Challenges. A case of the Namugongo Martyrs Shrines in Uganda. *Turystyka Kulturowa* 7: 117–31.

Norman, Alex. 2011. Celebrity push, celebrity pull: Understanding the role of the notable person in pilgrimage. *Australian Religion Studies Review* 24: 317–41. [CrossRef]

Onwezen, Marleen. 2018. Including context in consumer segmentation: A literature overview shows the what, why, and how. In *Methods in Consumer Research*. Sawston: Woodhead Publishing, vol. 1, pp. 383–400. [CrossRef]

Özdemir, Bahattin, and Faruk Seyitoğlu. 2017. A conceptual study of gastronomical quests of tourists: Authenticity or safety and comfort? *Tourism Management Perspectives* 23: 1–7. [CrossRef]

Prentice, Richard, and Vivien Andersen. 2003. Festival as creative destination. *Annals of Tourism Research* 30: 7–30. [CrossRef]

Rowley, Jennifer, and Catrin Williams. 2008. The impact of brand sponsorship of music festivals. *Marketing Intelligence & Planning* 26: 781–92.

Ruiz, Elena Cruz, Elena Ruiz Romero De la Cruz, and Francisco J. Calderón Vázquez. 2019. Sustainable Tourism and Residents' Perception towards the Brand: The Case of Malaga Spain). *Sustainability* 11: 292. [CrossRef]

Saayman, Melville, and Andrea Saayman. 2006. Does the location of arts festivals matter for the economic impact? *Papers in Regional Science* 85: 569–84. [CrossRef]

Shimazono, Susumu. 1998. The commercialization of the sacred: The structural evolution of religious communities in Japan. *Social Science Japan Journal* 1: 181–98. [CrossRef]

Shiomura, Takashi. 2009. Cultural polarization in increasingly nonlocal societies. *Sociological Theory and Methods* 24: 95–108.

Shuo, Yeh Sam Shih, Chris Ryan, and GeMaggie Liu. 2009. Taoism, temples and tourists: The case of Mazu pilgrimage tourism. *Tourism Management* 30: 581–88. [CrossRef]

Swatos, William H. 2006. *On the Road to Being There: Studies in Pilgrimage and Tourism in Late Modernity*. Leiden: Brill Academic Pub, vol. 12.

Takele, Yezihalemâ Sisayâ. 2019. International Tourism demand and determinant factor analysis in Ethiopia. *International Journal of Systems and Society IJSS)* 6: 27–51. [CrossRef]

Tanaka, Atsushi. 2014. Proposal of alleviative method of community analysis with overlapping nodes. *IEEE Fourth International Conference on Big Data and Cloud Computing*, 371–77. [CrossRef]

Timoshenko, Artem, and John R. Hauser. 2019. Identifying customer needs from user-generated content. *Marketing Science* 38: 1–20. [CrossRef]

Tomochi, Masaki. 2010. A model of a nested small-world network. *Sociological Theory and Methods* 25: 19–29.

Tsao, Karen. 2014. Mazu Cultural Festival and City Sustainable Development in Taichung. Available online: http://www.agenda21culture.net/sites/default/files/files/good_practices/taichung-eng.pdf (accessed on 1 August 2020).

United Nations Educational, Scientific and Culture Organization (UNESCO), and The Mazu Belief and Customs. 2009. Intangible Heritage Lists of 2009. Available online: http://ich.unesco.org/en/RL/mazu-belief-and-customs-00227 (accessed on 5 August 2020).

Viljoen, Armand, Martinette Kruger, and Melville Saayman. 2017. The 3-S typology of South African culinary festival visitors. *International Journal of Contemporary Hospitality Management* 29: 1560–79. [CrossRef]

Wang, Kuo-Yan. 2014a. Religious Service development of folk religion temples in Taiwan: A comprehensive perspective. *Review of Integrative Business and Economics Research* 4: 49–54.

Wang, Kuo-Yan. 2014b. When a Taoist temple serves as a seller and believer becomes a buyer. *Review of Religious Research* 56: 341–42. [CrossRef]

Wang, Kuo-Yan. 2019. Balancing environmental protection and folk belief: Surveying the attitudes of local worshipers. *DEStech Transactions on Environment, Energy and Earth Sciences PEEMS)*. [CrossRef]

Wang, Yue, Daniel Y. Mo, and Mitchell M. Tseng. 2018. Mapping customer needs to design parameters in the front end of product design by applying deep learning. *CIRP Annals* 67: 145–48. [CrossRef]

Wong, Maggie Hiufu. 2013. Mazu Mania: Free Food, Great Parties. Wait, This Is a Religious Festival? *CNN Travel*, April 12. Available online: http://travel.cnn.com/taiwan-mazu-religious-festival-pilgrimage-226351/ (accessed on 5 August 2020).

Woosnam, Kyle M., Kayode D. Aleshinloye, Manuel Alecto Ribeiro, Dimitrios Stylidis, Jingxian Jiang, and Emrullah Erul. 2018. Social determinants of place attachment at a World Heritage Site. *Tourism Management* 67: 139–46. [CrossRef]

Ying, Tianyu, and Jun Wen. 2019. Exploring the male Chinese tourists' motivation for commercial sex when travelling overseas: Scale construction and validation. *Tourism Management* 70: 479–90. [CrossRef]

Zaidman, Nurit. 2003. Commercialization of religious objects: A comparison between traditional and new age religions. *Social Compass* 50: 345–60. [CrossRef]

Zhang, Carol Xiaoyue, Lawrence Hoc Nang Fong, Shi Na Li, and Tuan Phong Ly. 2019. National identity and cultural festivals in postcolonial destinations. *Tourism Management* 73: 94–104. [CrossRef]

religions MDPI

Article

Buddhist Pilgrimage and the Ritual Ecology of Sacred Sites in the Indo-Gangetic Region

David Geary [1] and Kiran Shinde [2,*]

[1] Department of Community, Culture and Global Studies, University of British Columbia, Kelowna, BC V1V 1V7, Canada; david.geary@ubc.ca
[2] Department of Social Enquiry, School of Humanities, La Trobe University, Melbourne 3083, Australia
* Correspondence: K.Shinde@latrobe.edu.au

Abstract: In contemporary India and Nepal, Buddhist pilgrimage spaces constitute a ritual ecology. Not only is pilgrimage a form of ritual practice that is central to placemaking and the construction of a Buddhist sacred geography, but the actions of religious adherents at sacred centers also involve a rich and diverse set of ritual observances and performances. Drawing on ethnographic research, this paper examines how the material and corporeal aspects of Buddhist ritual contribute to the distinctive religious sense of place that reinforce the memory of the Buddha's life and the historical ties to the Indian subcontinent. It is found that at most Buddhist sites, pilgrim groups mostly travel with their own monks, nuns, and guides from their respective countries who facilitate devotion and reside in the monasteries and guest houses affiliated with their national community. Despite the differences across national, cultural–linguistic, and sectarian lines, the ritual practices associated with pilgrimage speak to certain patterns of religious motivation and behavior that contribute to a sense of shared identity that plays an important role in how Buddhists imagine themselves as part of a translocal religion in a globalizing age.

Keywords: Buddhist pilgrimage; ritual ecology; Buddhism; Bodhgaya; Buddhist heritage

Citation: Geary, David, and Kiran Shinde. 2021. Buddhist Pilgrimage and the Ritual Ecology of Sacred Sites in the Indo-Gangetic Region. *Religions* 12: 385. https://doi.org/10.3390/rel12060385

Academic Editor: Claude Stulting

Received: 21 April 2021
Accepted: 20 May 2021
Published: 26 May 2021

1. Introduction

According to early Buddhist texts such as the Pali Canon, the birthplace of Buddhism is associated with the geographic region known as *Majjhimadesha* ("middle country"). Sometime between the sixth and forth centuries BCE, it is believed that the Buddha lived and taught in the foothills of Himalayas in the low plain lands of the Terai and Indo-Gangetic region that extends east to what is now the Indian states of Bihar and West Bengal. Although the transregional histories and ties to the Buddha-dharma in the Indian subcontinent spans over numerous centuries, the modern rebirth and renewal of this ancient pilgrimage geography is a relatively recent phenomenon spanning the late nineteenth and twentieth century (Geary and Mukherjee 2017; Ray 2014; Sharma 2018; Singh 2010). Today, a journey through Buddha's homeland is a dream for many Buddhist followers and for those who seek the path of Awakening (*bodhi*) as exemplified by the Buddha's life. In this journey, four places have been elevated as part of a sacred geography due their association with important biographical events in the life of the religious founder. These places are: Lumbinī, (birthplace of Buddha, Nepal); Bodhgaya (the place of Buddha's enlightenment), Sarnath (where Buddha delivered his first sermon), and Kuśinagarī (where Buddha took nirvana). Although not limited to these sights, these places have become sacred foci for the convergence of multiple expressions of Buddhist devotion that cut across national, cultural–linguistic, and sectarian lines. Collectively and individually, these sacred sites are visited by hundreds of thousands of devotees and nondevotees that are drawn to the "spiritual magnetism" (Preston 1992) of place but also the melding of Buddhist teachings, culture, and heritage values as popular religious and tourist spaces.

Many recent studies have deliberated about the meaning of such pilgrimage for those who visit these iconic Buddhist sites (Bruntz and Schedneck 2020; Geary 2014; Hall 2006; Huber 2008). A few scholars have positioned the importance of social engagement wedded to soteriological aspirations as one of the highest forms of meaning that life-long followers seek in their dwelling in Buddhist places (Goldberg 2013; Kitiarsa 2010). Others have discussed forms of contestation that is generated by those interested in touristic aspects that can be seen as contradicting the peace and serenity that is sought after among some Buddhists and spiritual tourists themselves (Geary 2008; Hung 2015; Hung et al. 2017; Philp and Mercer 1999). Such seeming contradictions are played out in the same spatial environment, not to mention the diverse national and sectarian expressions of Buddhism, including its own local adaptations among new Buddhist converts and communities. However, only a handful of studies offer some insights on the specific religious performances, rituals, and "material affordances" (Gibson 1977) in Buddhist pilgrimage sites and how these religious activities both inscribe and produce certain meanings (Seelananda 2010; Wong et al. 2013).

This paper aims to describe the ritual ecology related to the main Buddhist pilgrimage sites in North India and Nepal, showing how the material and corporeal aspects of Buddhist religious experience contribute to the distinctive religious sense of places, as well as some of the unique pressures on the built environment as spaces of religious heritage. Our use of the term ritual ecology builds on the definition of a religious setting provided by Kerestetzi (2018, para 2) as "a coherent space in which objects, bodies, actions, and ideas form a system—or even an ecosystem" that enables us to dispense with some of the dualistic approaches such as faith/matter and sacred/profane. Although much of our descriptions of ritual phenomena can be generalized across several main Buddhist pilgrimage sites, our analysis is greatly informed by sustained fieldwork at a few prominent sites, especially Bodhgaya—the seat of enlightenment located in south Bihar. Thus, a study of ritual ecosystems at Buddhist sites, such as Bodhgaya, has the potential to provide important insights into how the vitality and vibrancy of a site for Buddhist practitioners and visitors can be more effectively managed by state governments, archaeological and heritage agencies, and religious authorities. A focus on rituals as a mode of analysis also helps to make sense of the corporeal and cosmic synthesis of Buddhist world-making and the relationship between spatiality and materiality that is so important to the emplacement of a sacred geography (Shinde 2020).

It is important to keep in mind that this paper is a preliminary effort to contextualize the predominant forms of religious practice and how that gives rise to certain patterns of sacralization rather than an exhaustive all-encompassing picture of the various rituals and performances undertaken by the different international and local Buddhist groups. Due to the absence of normalized ritual protocols and diverse histories and lineages, there is considerable variability and innovation across the Buddhist spectrum. The remainder of the paper is divided into the following sections: In the next section, we provide an overview of Buddhist pilgrimage drawing on recent scholarship and translations. Following this, we provide a more comprehensive taxonomy of prominent Buddhist rituals and performances in sacred sites with a specific reference to Buddhist sites in North India and Nepal. In the final section, we highlight some main themes deriving from the research and discuss how the influence of rituals and forms of spatial management impact the religious environment, the pilgrimage-tourism economy, and heritage values that contribute to the sociocultural fabric of Buddhist sites.

2. Buddhist Pilgrimage and Paths of Convergence

Although distinctive forms of Buddhist pilgrimage and religious travel are common in various parts of Asia, such as Thailand, China, Taiwan, Japan, and so forth, (Bruntz and Schedneck 2020; Choe and O'Regan 2015; Mason and Chung 2018; Wong et al. 2016), our emphasis here is on the paths of convergence associated with the biography of the founder and how this core Indic geography absorbs and attracts pilgrims from multiple

Buddhist traditions (as well as non-Buddhists). What is the religious and historical basis for Buddhist pilgrimage in the Indian subcontinent? What are the main motivations and goals for undertaking Buddhist pilgrimage, and how is that expressed through religious journeys and the ritual ecology of sacred sites in India and Nepal?

What is usually taken as the "locus classicus" (Huber 2008) of Buddhist pilgrimage are the four main sites—Lumbinī, Bodhgaya, Sarnath, and Kuśinagarī—recommended by the Buddha to visit after his death in the famous Mahāparinirvāṇa sūtra (MPNS). These four sites mark his birth, enlightenment, first sermon, and *parinirvāṇa* respectively. Counter to the usual reification of the four-site model of Buddhist pilgrimage, in a recent chapter, Strong (2014) scrutinizes some of the different translations of the famous discourse in Pali, Sanskrit, and Chinese, noting some noteworthy differences among them. Of particular importance is a preceding paragraph that is common in Chinese translations which he refers to as "Ananda's lament". In this passage, "Ananda expresses to the Buddha his sorrow that, in times past, after they had spent the rains-retreat in various places, mind-cultivating *(mano-bhāvanīya)* monks used to come for the sake of seeing (*dassanāya*) the Buddha (i.e., of having *darśan* with him), but that now, after the Blessed One's demise, they will no longer be able to do so. And this grieves him" (Strong 2014, p. 51). Furthermore, Ananda specifies that "the monks used to journey not only in order to see the Buddha, but to worship him and make offerings, to ask questions and to receive instruction on the Dharma. Once the Blessed One enters *parinirvāṇa*, there will no longer be any point in their coming" (Strong 2014, p. 51). This leads Ananda to pose the obvious question about the preservation of the Dharma and how the monks should respond to his pending absence. In response to Ananda's query, the Buddha appears to recommend the importance of undertaking pilgrimage to the four sites as a way to earn rebirth in heaven but also as a solution to his absence whereby followers may "continue to 'see' and venerate him after his passing" (Strong 2014, p. 51).

On the basis of his reading of Ananda's lament, Strong makes a number of observations that speak to the importance of pilgrimage and ritual in the Buddhist tradition. Firstly, the passage cited above hints at what may have been an earlier pattern of pilgrimage that existed during the lifetime of the Buddha and suggests the importance of several places located on the margins of urban centers that correspond with the Buddha's rain retreats, such as Vaiśālī, Rājagṛha, Śrāvastī, and Saṃkāśya. Although these four locations are frequently described as "secondary sites" that evolved out of the primary "locus classicus" prescribed by the Buddha to Ananda, as Strong suggests, the reverse may actually be the case when referring to the origins of pilgrimage. This tradition is also corroborated by several passages in the Pali Vinaya as well as Buddhaghosa who describes the importance of monks and nuns, especially those who "cultivate the mind" (*mano-bhāvanīya*) undertaking pilgrimage to visit the Buddha at the end of the rains retreat as a form of veneration and a means of receiving instruction. Another important distinction here is how the Buddha's recommendation of the four sites that follows Ananda's lament is no longer directed toward "meditating monks and nuns" but rather encompasses all Buddhists who wish to journey and see the Buddha, including laymen and laywomen (Strong 2014, p. 52). In other words, as Strong writes, "If pilgrimage originally started as a strictly monastic tradition for meditating monks during the lifetime of the Buddha, it would seem that, with the Buddha's passing, it became more generally a devotional tradition for all Buddhists" (Strong 2014, p. 52). Although more speculative, another reason for the displacement of the earlier four sites associated with the rain retreats may have been the desire to build a distinct biographical tradition that would not only be open to wider Buddhist publics but could be distinguished from other popular and competing religious movements at the time, such as Jain centers of pilgrimage that were also present in places such as Rājagṛha, Śrāvastī, and Vaiśālī (Strong 2014).

When we turn to the goal of undertaking pilgrimage to the four sites as expressed by the Buddha to his companion Ananda, according to Strong, two key Indic concepts need to be highlighted. Firstly, by foregrounding the Buddha-*carita* or Buddha-biographical tradition, which focuses on the activities (*carita*) performed by Sakyamuni Buddha, and the places associated with these acts, this provides an opportunity for Buddhist devotees to undertake a form of "*darśan* by proxy"—even after his death and *parinirvāṇa*. Another way of viewing *darśan* by proxy, "is to see it as *darśan* (direct seeing) on its way to being *anusmṛti* (vivid imaginative or meditative recall). This is important because, in the non-Pali versions of the MPNS, the language of 'seeing' disappears, and is replaced by the language of 'recall'" (Strong 2014, p. 56). Given the emphasis on imaginative and meditative recall, according to André Bareau, this could imply that instead of undertaking pilgrimage to the physical sites "meditating on moments of the Buddha's life for the purpose of recalling him (buddha-*anusmṛti*)" and his different aspects of Buddhahood might be sufficient as alternative to seeing (*darśan*) and also indicate that this could "be done anywhere at any time" (cited in Strong 2014, p. 57). It is also worth noting that the emphasis on *anusmṛti* would have also likely been more integral to the aspirations of monks and nuns rather than laypeople.

The second key Indic concept that needs to be highlighted with reference to pilgrimage practices is the importance of sacred places as conduits for arousing *saṃvega*. In his translation of the concept, Strong writes "*Saṃvega* is a complex notion, but it is generally thought to be a religious emotion in the face of the sight of the truth of impermanence or any other feature of *saṃsāra* that gives rise to a desire to adopt the religious life" (Strong 2014, p. 54). Although the Buddha does not explicitly address why these four pilgrimage sites result in *saṃvega*, it would imply, according to Strong that

> *pilgrimage to (and so darśan at) these sites entails a dual and contradictory emotion of remembering the presence of the Buddha at a particular time and place of his life by 'seeing' him, and of being moved by a realization of his present absence there. In this sense, pilgrimage sites would be akin to relics which, I have argued elsewhere, are essentially reminders of a biographical process that affirms both the presence and the absence of the departed one.* (p. 54)

Although it is tempting to suggest that there would have been some clear benefits to visiting the four sites in a biochronological order and to retracing his life journey, there does not appear to be any evidence of this from pilgrim records (Strong 2014). The only literary instance of this, as Strong points out, is the famous Aśokāvadāna legend where he undertakes pilgrimage and marks the sites with *caityas*.[1]

What is evident in these translations of the MPNS are the ways in which Buddhist pilgrimage can be defined as the "ritual re-enactment of religious narratives" (Castelli 2003 cited in Deeg 2014, p. 9) that has empowered a sacred landscape and is representative of a "concentrated evocation of the Tathagata's [Buddha's] life" (Coleman and Elsner 1995, p. 195). An important feature of Buddhist pilgrimage is an emphasis on rebirth. The Pali version of the Buddha's discourse mentions that those followers who die during their pilgrimage to the four sites,[2] or after having made it, "will by virtue of their merit be reborn in heaven (or in the favourable human state)" (Strong 2014, p. 59). This sentiment continues today, especially among lay Buddhists, and that is its close association with accruing merit on the journey, or on the journey, in this life and for those to come. Thus, pilgrimage in the Buddhist tradition has taken on an important soteriological dimension as an external metaphor of an internal journey toward spiritual awakening (Coleman and Elsner 1995).

However, it is also important to clarify that pilgrimage is not only a process of embodied recollection that allows one to internalize the teaching of the dhamma but also involves contact with sacred traces and various material objects, such as relics, that play an important role in sanctifying space and representing the physical aspects of the narrative.[3] As vital entities of the Buddha's present absence, the veneration of relics through the treatment of his corporeal remains and the subsequent construction of Buddhist *stūpas* or *caityas* may well be some of the oldest forms of Buddhist pilgrimage (Deeg 2014). Through their division and multiplication, relics have also played an important role in the geographic and ever-expanding repertoire of stories associated with the Buddha's life through the layering of various hagiographical legends and events. Although Buddhist relics take on a particular significance in the Buddhist homeland given their association with the Buddha's biography and/or other enlightened persons, as Maud (2017, p. 422) writes, "the geography of relics is not static but continually evolving, and the movement, exchange, and relocation of relics index the transformations of Buddhism's presence in the world, its geographical extension, and the fluctuating fortunes of its adherents".

While much has been written on the spread of bodhi tree offshoots (Maitland 2018; Nugteren 1995; Ober 2019), the replication and reproduction of Mahabodhi temples (Asher 2012; Guy 1991), and the afterlives of relics (Mukherjee 2018; Sharf 1999; Trainor 1997) that all contribute to a "shifting terrain of the Buddha" (Huber 2008), there is no denying that these material objects and their circulations have become important foci of ritual veneration (and various sociopolitical power struggles) that mediate the Buddha's presence and are central to accruing merit. Due to their heightened status in relation to a sacred place, they have also been instrumental in the material prosperity of certain localities, including the development of monasteries around sacred sites by attracting donations and gifts from various lay pilgrims and patrons. As Schopen (2004, p. 100) writes, "relics gave rise to festivals; festivals gave rise to trade; trade gave rise to gifts and donations" (cited in Maud 2017, p. 424). Donation inscriptions as well as contemporary ritual practices at various sacred sites point to a long history of gifts and offerings to the local *saṅgha* or the *stūpa* itself (Deeg 2014, p. 10). Through the actions of various pious rulers as well, at least since the time of Asoka, we also see how patronage translates into material transformations such as various infrastructural improvements, the building of rest houses, and improved connectivity (Geary 2017).

Thus, by undertaking pilgrimage for the devout, not only does this entail a journey to see the Buddha and arouse certain religious emotions associated with his worldly activity and subsequent material traces, but it is capable of inspiring one along the path of Buddhadharma by embodying the sacred paradigm of the religion's founder. Together, relics and places of pilgrimage serve as mnemonic devices and spaces of memorialization that both embed and diffuse the spiritual biography of the Buddha, as well as many of his faithful disciples. In the next section, we provide a survey of some of the dominant ritual forms that are visible at Buddhist pilgrimage sites in North India and Nepal today.

3. Sacralizing Space through Ritual Performance

Although Buddhist pilgrimage is a diverse phenomenon that cross-cuts many different rituals and motivations, based on the analysis above, especially the recent translations by John Strong, a few key patterns can be discerned from both historical and contemporary practices. It could be argued that the primary motivation to undertake pilgrimage is: *a journey to see and/or recall the Buddha and be moved by a realization of his present absence*. In Buddhism, unlike in other religions, such as Islam, it is not necessary or obligatory to undertake pilgrimage or perform certain rites of passage that are related to the fundamental religious tenets. Group composition, length of the pilgrimage journey, and the means of transportation also vary significantly across Buddhist cultural groups. For example, among some Buddhist communities, such as Tibetans, undertaking the pilgrimage journey—*gnas skor ba* (Trans. "going around a [sacred] place")—by foot and sometimes over several years and months is seen to be more meritorious than expediated modes of modern transportation such as airlines and buses. Despite these variations in length and mode of travel, pilgrimage, one could argue, is universally seen as an auspicious opportunity to venerate the sacred traces and through various ritual practices accrue religious merit (*punya*) that hopefully provides certain soteriological advances such as a favorable rebirth and ultimately liberation. However, as we illustrate below, undertaking pilgrimage and accumulating good karma may also be directed toward more mundane this-worldly aspirations and rewards that have an important social and economic role in society.

Once at a Buddhist sacred site, what activities do pilgrims undertake to fulfil the goals and aspirations of the journey? From this wider normative definition of Buddhist pilgrimage, we can now turn to some of the prevalent ritual forms and expressions that take place at sacred sites in India and Nepal. In creating this ritual taxonomy, it is important to keep in mind that these categories are not mutually exclusive, and there is considerable overlap between the different practices in term of their desired goals and intention. For example, it could be argued that merit-making is infused within each of the ritual categories below. We also do not focus on more esoteric forms of ritual engagement such as secret initiations that may take place between a discipline and guru at a sacred site. Rather, our focus is on public displays of ritual that mediate Buddhist pilgrimage and the sacred built environment.

3.1. Calendrical Rites and Memorial Ceremonies

These are rituals that occur at regular intervals throughout the Gregorian and lunisolar calendar that mark a date or event of symbolic significance for the Buddhist community. Below are some of the more prominent ritual events.

Vesak. (S. Vaiśākha; P. Vesākha; T. Sa ga zla ba): Based on the ceremonial cycle of Buddhist holidays, especially following the Theravada and Tibetan tradition, the foremost calendrical ritual is that of Vesak (also referred to in vernacular terms as Buddha Purnima or Sawa Dawa in Tibetan). This full-moon day falls on the fourth lunar month of the traditional Indian calendar (usually corresponding in April–May) and commemorates three important events in the life of the Buddha: his birth, awakening, and *parinirvāṇa*. Although widely celebrated in many parts of Asia, especially in South East Asia, it takes on particular significance in the holy land and is frequently combined with other meritorious actions (see below). On this occasion, numerous functions and rituals are arranged to honor the memory of the Buddha, such as elaborate processions through the surrounding town toward the main temple or stupa (Figure 1). This event is usually organized and planned by local monastic councils and temple committees. Although the date of Vesak has become standardized in contemporary India and in many ways has become a symbol of unification for the different monastic groups and lineages in Buddhist sacred places, it is important to keep in mind that the actual date and symbolic importance of Vesak varies considerably amongst Buddhist communities.[4]

Figure 1. Procession and re-enactment of events in the Buddha's life by school children during Vesak in Bodhgaya. (Photo from the author).

Kathina Dana: This refers to the "great robe offering" ceremony in the Theravada tradition that marks the end of the *vassa* ("rains") retreats and is celebrated with many *sanghadana's* held between different local monasteries and temples, especially from South and South East Asia (Figure 2). On these occasions, lay pilgrims visit members of the sangha and offer cotton cloth for the purpose of making new robes, gifts, feasts, and donations to the temple to support the monks and nuns in their pursuit.

Figure 2. Offerings by lay pilgrims during the Kathina Dana at a nearby temple in Bodhgaya. (Photo from the author).

Other collective gatherings: It is common for Buddhist communities to organize ritual gatherings at sacred sites to celebrate lunisolar events throughout the year. Losar (T., lo-gsar), the Tibetan New Year, is usually celebrated around February (according to the Tibetan Calendar) and lasts 3–15 days. In July, Theravada and Tibetan Buddhists in Sarnath celebrate *Asalha* "Dharma Day" or *Chökor Düchen* (T. chos 'khor chen), which honors the day the Buddha first taught the four noble truths and turned the wheel of dharma on the full-moon night of the month of Asalha. Magha Puja—a time when 1250 Buddhists came together to pay their respect to the Buddha—has now become a festive occasion among Buddhists from South East Asia to honor the Sangha and provides opportunities for lay and monastic people to reaffirm their commitment to the Buddha-dhamma. Many Burmese groups celebrate the *Abhidhamma* day to mark the day the Buddha went to the Tushita heaven to teach his mother. On the full-moon day in March, Buddhists from the Mahayana traditions of Tibet and East Asia celebrate Avalokitsevara's birthday—one of the eight great Bodhisattvas. Rituals and forms of worship on this day involve the recitation of the mantra *'om mani padme hum'* and the active practice of compassion (*karuna*) that exemplifies the Bodhisattva ideal to save and protect beings. In the spirit of the Bodhisattva ideal, among Chinese Buddhists, the Water Land Dharma Function is also prominent in some sacred sites and is presided over by high-ranking monks who invite beings of higher realms to ease the suffering of those below. Throughout each month there are also important observances such as *uposadha* (P. uposatha; T. gso sbyong) that follow the new and full moon. On these occasions, ordained members of the sangha assemble together and intensify their practice of the Buddha's teachings by making a conscious effort to renew their commitments to moral conduct through their recitation of the *bhikkhu-pratimoksa* from the *Vinaya Pitaka* (monastic rules) as well as undertake various forms of self-cultivation and mental purification such as meditation. Among laypeople, it is important to maintain the Eight Precepts (*astahga-sila*).

In addition to these popular Buddhist ritual events that correlate with the lunisolar calendar, there are increasingly a number of memorial events that recognize important historical figures, teachers, and Rinpoches. Some of the more prominent events that we have witnessed include the birthday celebrations of the Dalai Lama on 6 July, the birth of the Thai King on 28 July, and the birth and passing of Anagarika Dharmapala (17 September and 29 April). Among new Indian Buddhists, the birth of Bhimrao Ramji Ambedkar on 14 April, his conversation on 14 October, and his passing on 8 December are particularly auspicious and may also correlate with conversion ceremonies (see below).

3.2. Sacred Movement, Meditation, and Veneration of Objects

These embodied rituals reflect both individual and collective forms of aspiration and spiritual cultivation that are directed toward both soteriological and mundane goals. This category of rituals is perhaps the most dominant and expressive modality that can be observed at Buddhist pilgrimage sites. As Wallace writes "all formal practices of focused concentration, mindfulness, prayer, chanting, and other ritual activities are performed as a means of cultivating one's heart and mind or expressing one's faith" (Wallace 2002, p. 36). Similarly, Huber (2008, p. 310) shows in his study of Tibetan pilgrimage, the ritual work "performed by the pilgrim's body is understood as 'cleansing' or removing certain types of embodied moral and cognitive defilements that hinder progress toward salvation". Although there is a wide spectrum of veneration rituals and practices involving sacred movement that overlaps with the next section, we have chosen to highlight four here:

Pradaksina (P. padakkhina; T. skor ba) is the common practice of showing reverence by circumambulating a holy object, person, or place. Among Tibetans, this is referred to as *kora, kor* meaning circle in Tibetan. In English, making *kora* is the same as undertaking circumambulation, from the Latin *circum* (around) and *ambulare* (to walk), and in many ways, it represents a microcosm for the pilgrimage journey itself. To circumambulate as a form of religious veneration involves making a clear and conscious connection with something that is regarded as special and holds sacred power (Figure 3). As Keown writes "Locating the encircled object at the centre symbolizes its centrality in the lives of those who walk around it. The activity also represents integrity and cosmic harmony in mirroring natural phenomena such as the clockwise course the sun was believed to follow over the surface of the earth" (2004, p. 218). At Buddhist pilgrimage sites, the most common objects of devotion are stupas or temples but might also include relics, bodhi trees, pillars, images, monasteries, shrines, and other natural features of a sacred landscape such as mountains and hilltops. Circling a stupa or sacred object three times can also be interpreted as a way of showing respect for the Triple Gem. In places such as Bodhgaya, it is not uncommon to see a steady stream of Himalayan Buddhists moving clockwise in large numbers around the upper and lower circumambulation path swirling their prayer wheels, muttering guru-mantra recitations, and counting their mala prayer-beads on their fingertips throughout the day. For many of these Tibetan Buddhists, it is believed that performing *kora* while counting prayers or mantra and/or spinning a prayer wheel with a strong motivation helps to multiply the merit.[5]

Figure 3. Tibetan nun with prayer wheel outside Mahabodhi Temple. (Photo from the author).

Prostration: (P. panipāta, S. namas-kara) This is a common gesture of deep respect used among Buddhists to show reverence to the Triple Gem (three times) and a common practice of veneration at sacred sites that helps to accumulate merit. Although the type of prostration and accompanying prayers vary across Buddhist traditions (such as half and full length with emphasis on different verses), it is quite common in pilgrimage sites to see Buddhists prostrating along a circumambulation path or before a particular image, teacher, or stupa (Figure 4). In the Pali Canon, there are several *suttas* that mention laypersons prostrating before the living Buddha (Mills 1982). The action, when it involves dropping the entire body forward and stretching it full length on the ground, also requires considerable physical and mental effort. For these reasons, prostrations, like forms of meditation (see below), can be interpreted as a means of purifying one's body, speech, and mind of karmic defilements to further progress along the path to enlightenment.[6] For example, in the northwest corner of the Mahabodhi temple complex, during the winter season, it is common to see hundreds of maroon-clad monks performing full-length prostrations on long wooden boards in sets of a hundred thousand over several weeks and months in an effort to complete their preliminary practices.

Figure 4. Full-length prostration before the Mahabodhi temple, Bodhgaya. (Photo from the author).

Meditation: We use the English term meditation to refer to a range of more specific techniques and practices that express forms of restricted bodily movement that bring about physiological changes and encourage an altered state of consciousness which is amendable to spiritual development. The two most prominent forms of Buddhist mental cultivation that are used for concentration purposes and to focus the mind are calming meditation (*samatha*) and insight meditation (*vipasyana*). Although the various forms of meditation are not directly associated with pilgrimage rituals, they are increasingly prevalent at sacred sites and are suggested to play an important role, especially among Mahayana Buddhists, as means of enacting one's Buddha-nature that can also be interpreted as a form of veneration toward Shakyamuni and the Dharma. This is especially the case in Bodhgaya where pilgrims can be found throughout the Mahabodhi temple complex in quietude and stillness. Although it has been emphasized that these meditation techniques were rarely practiced by laity in the past, today, they figure prominently in the ritual ecology of sacred sites among individuals and groups as a means of removing mental impurities and cultivating techniques that speak to Buddhist aspirations of achieving Nirvana across a wide spectrum of practitioners.

Dance: although not nearly as common as the embodied rituals highlighted above, forms of religious dance, especially among Tibetan Buddhists, often take place at sacred sites and are often part of larger religious gatherings and festivals. Lama dancing (T. 'chams), in particular, are performed by masked dancers accompanied by ritual music and involve the enacting of various religious dramas that depict important Tibetan mythological themes such as the subjugation of demons.

3.3. Chanting, Recitation, and Prayer Festivals

Despite the popular association of Buddhism (and the Buddha) with forms of quietude and stillness, Buddhist pilgrimage sites are densely layered sonic environments. As part of the "ritual re-enactment of religious narratives" (Deeg 2014, p. 9), the individual and collective recitation of certain Buddhist teachings, discourses, and prayers play an important role in elevating the sanctity of the place while paying homage to the Buddha, Dharma, and Sangha. As part of this "grammar of sanctification" (Eck 2012), certain pilgrimage places have become important locales for particular Buddhist discourses and stories associated with the Buddha's life (lives) along with his disciples. One example is the Lotus and Heart Sutra in Vulture Peak, Rajgirha. Given the importance of the Lotus Sutra in Rajgir-Nalanda for Mahayana Buddhists, especially the Nichiren school of Buddhism, it is common to see gatherings of Japanese Buddhists practicing the *daimoku*, chanting the title of the sutra in the form of "*namo myoho renge hyo*" (Jap.) a phrase meaning "Hail to the Scripture of the Lotus of the Wonderful Dharma", which is itself an expression of devotion to the Lotus Sutra's truth and power. Among Tibetan Buddhists, in most pilgrimage places, one hears the words "*om mani padme hum*". Although the exact translation of this verse has a complex grammatical structure and is not easily amendable to a definitive translation, it does gesture toward the meaning: "homage [or Praise] to the Jewel-Lotus One". This popular mantra is widely associated with the venerated Bodhisattva of compassion Avalokitsevara and is frequently placed in prayer wheels. Other common prayers include the refuge prayer (Kyabdro in Tibetan), which is the act of taking refuge in the Three Jewels. In the ancient Pali language, the chant goes,

Buddham saranam gacchami. "I take refuge in the Buddha".
Dhammam saranam gacchami. "I take refuge in the Dharma".
Sangham saranam gacchami. "I take refuge in the Sangha".

Accompanying these prayers and mantras is the use of a *mala* (skt)—a form of rosary—held in the right hand and fingered by pilgrims to keep count of the number of recitations made during their worship. Although the string of beads on a mala can vary in number and length, the most common number is 108.[7] In terms of the type of bead, increasingly they are made from sandalwood or seeds of the Bodhi Tree under which the Buddha gained enlightenment, although rosaries can also be made from a range of other materials such as wood, hard-nut kernels, crystal, and bone (Keown 2004, p. 171).

Given the importance of Bodhgaya in terms of multiplying aspirations, Buddhist sacred sites, such as the Mahabodhi temple, have become important staging grounds for several *monlams* (or prayer festivals) that involve prayers, mantras, and long liturgies guided by high-ranking lamas and Rinpoches from the different Tibetan schools (Figure 5). Although many of the original *monlam* prayer festivals in Tibet commemorated the Buddha's defeat of the heretical teachers at Sravasti, at several Buddhist sites in India and Nepal, they are now conflated with a broader set of global discourse such as "world peace" (see Huber 2008). Over a period of ten days (sometimes more), the festival includes prayers performed three times each day at the sacred site and may be accompanied by tantric practices of visualization, rituals for expiation of misdeeds committed during the previous year, and a rededication to the principles of Buddhism for the coming year (Buswell and Lopez 2013, p. 831). As part of the materiality of devotion that accompanies these large ritual gatherings, elaborate altars are constructed on the side of the temple or beneath the Bodhi tree, arranged with silver water bowls, receptacles filled with rice, and elaborate multicolored *tormas* made of colored barley flour and butter that are used as an offering or propitiation to various spiritual beings (Figure 6).

Figure 5. Monlam event under the boughs of the Bodhi tree in Bodhgaya. (Photo from the author).

Figure 6. *Tormas* during the Monlam chanting in Bodhgaya. (Photo from the author).

In recent years, the Light of Buddhadharma Foundation (LBDFI) founded by Wangmo and Richard Dixey (the former is the daughter of Tarthang Tulku) have taken a leading role in organizing the annual International Tipitaka chanting ceremony. The chanting ceremony brings together upwards of 3500 Theravada monks, nuns, and lay followers from ten South East and South Asia countries to chant the teachings of the Buddha, as recorded in the Pali Canon. Each country is represented by separate national stalls and simultaneously chants over one another during the scheduled event (Figure 7).

Figure 7. Thai participants in the International Tipitika Chanting ceremony in Bodhgaya. (Photo from the author).

3.4. Spiritual Teachings, Initiation, Counsel, and Retreat

Many Buddhists undertake pilgrimage to gain access to high-ranking monks or nuns and their teachings at a sacred site. Contact with a guru or members of the sangha may translate into forms of private counsel, spiritual initiation, and blessings. Because the vast majority of pilgrim groups reside in monasteries located at each of the main pilgrimage sites, senior monks and nuns also play an important role as ritual mediators and guides for merit-making activities, providing dhamma talks that recount stories of the Buddha's life and leading groups in chanting and meditation.

Given the importance of spiritual counsel, several Buddhist sites have evolved into important centers of study (of Buddhist philosophy and practice) and retreat (meditation) that draw both secular and religious participants. For example, there are several Goenka-inspired Vipassana Centres on the outskirts of various Buddhist pilgrimage sites attracting various participants throughout the year. There is a six month spiritual program offered by Root Institute for Cultural Wisdom in Bodhgaya. Through the peak months of December and January, many of the pilgrims from the Himalayan regions migrate to Bodhgaya, escaping the cooler climates by following respected lamas or Rinpoches from the various Tibetan schools and lineages (Geary 2014). This affords unique opportunities to attend various teachings, such as Ayang Rinpoche's annual phowa (T. 'pho ba) course held in Bodhgaya each year. There are also increasingly growing numbers of educational groups and pilgrimage tours such as the American-based Carleton (formerly Antioch) Buddhist Studies Program that involves both contemplative and educational university accredited courses that take place in residence at the Burmese *vihara*.

3.5. Dana and Making Merit: Donations, Offerings, and Service

Dana (Skt.; Pali) is loosely translated in English as "generosity" and is a key Buddhist virtue and source of great merit (*punya*) that is believed to be multiplied at a sacred site. As the first of ten perfections in the Pali tradition, among Theravada Buddhists, dana is seen as altruistic behavior that is instrumental in overcoming selfishness and attachment. In the Mahayana tradition, it usually refers to the Perfection of Generosity (Keown 2004, p. 69). Among all Buddhists, practicing different forms of generosity and making offerings provides certain soteriological advantages such as achieving a better rebirth. Offerings at a sacred site may also be directed toward more mundane material benefits such as the fulfillment of a special vow and/or to honor a deity in return for financial prosperity and good health. However, what defines a worthy recipient of a gift, also known as a 'field of merit' (*punya-ksetra*) is not always clear and may vary considerably across Buddhist cultural groups and among lay and monastic members. First and foremost, at a pilgrimage place, a valued field of merit is likely to include the giving of alms to members of the ordained sangha, making offerings toward a buddha image or shrine, the refurbishment, or embellishing of a temple or stupa, and the financial support of religious infrastructure and construction works such as a *dharamshala* (pilgrim-lodge) for pilgrims. As several scholars have shown in recent studies, increasingly, sacred sites have become launching pads for spectacular "materializations of merit" (Askew 2007; McDaniel 2016) that may also be used to compete for pilgrim and tourist patronage. In contrast to these spectacular displays, what is more common among lay pilgrims are ephemeral ritual offerings that are tied to the efficacy of karma and are part of the everyday materiality of devotion such as sweet perfumes, flowers, attaching prayer flags, providing bowls of fruit and foodstuff, water, oil lamps, candles, and incense. Applying gold leaf on sacred objects and images such as stone footprints (to the chagrin of conservation authorities) is also a popular practice among some South East Asian Buddhist groups. In a recent chapter by anthropologist Cook (2018), she accompanied a group of 120 Buddhist pilgrims from Chiang Mai, northern Thailand, to Nepal and India to make the journey around the four sites. In one instance, Cook describes the ways in which Thai lay pilgrims internalize and actualize the truths of suffering and impermanence in Kuśinagarī by symbolically cremating the names of deceased family members on small slips of paper, which allow them to create a tangible connection between the death of the Buddha and their own personal experiences of loss. These actions are not only morally transformative to the participant, but they also infuse and enhance the sacred power associated with the Buddhist sites because "the more people share in a meritorious act, the more merit is created" (Cook 2018, p. 45).

Another meritorious activity that is of great importance in Bodhgaya, especially to Theravada Buddhists, is the dressing of the gold-gilded central image of Buddha Shakyamuni with robes. During particularly auspicious days on the lunar calendar, the Buddha image is repeatedly dressed throughout the day by the temple management staff at the request of the devotee. The sanctification of robes by the Buddha is not only a great source of merit but also provides an important source of revenue for the temple management committee who later return the robes to the entrusted devotee, usually in exchange for a financial donation to the temple (Figure 8).

In several Buddhist sites in India and Nepal, making Dana may also translate into various forms of relief work and charitable projects that may fall under the umbrella term "socially engaged Buddhism" (Queen 2012), which involves applying Buddhist solutions to reduce various forms of suffering. The engaged Buddhist movement cuts across the lay–monastic divide and includes Buddhists from traditional Buddhist countries as well as more recent Western converts. Due to the poverty surrounding many Buddhist pilgrimage sites in the Indo-Gangetic region, this has given rise to a number of charitable initiatives that include supporting or sponsoring children at schools and/or orphanages, establishing an NGO, giving to beggars, and/or volunteer service to the community and environment with various initiatives ranging from health clinics to the building of public toilets. For some, this form of "engaged Buddhism" is an expression of the highest form of seeking (Goldberg 2013).

Figure 8. Offering robes to the main Buddha image in the Mahabodhi temple. (Photo from the author).

As we see in this section, providing Dana and accruing merit can be expressed in a range of ritual forms, from the recitation of certain verses, offerings to the sangha, donations, and social work to name a few. Although there are multiple modalities for the transference of merit, one could argue that undertaking pilgrimage itself represents one of the highest forms of merit-making in the Buddhist world given the ways in which other ritual practices are enfolded within the journey. For these reasons, pilgrimage places are very auspicious locales for undertaking virtuous deeds which are akin to a merit multiplier effect.

3.6. Temporary Ordinations and "Conversion" Ceremonies

Although not nearly as common as some of the ritual behavior described above, pilgrimage places are also valued spaces for deepening ones commitment to the Triple Gems through forms of temporary ordination (*P. pabbajja*—"to go forth") as well as conversion. In terms of temporary ordination, this is especially the case for many Theravada Buddhists, especially from Thailand, where temporary ordinations are like a rite of passage for young men in Thai Buddhist society who are expected to spend some time in a *vihara*. Although it was not practiced during the Buddha's time and it is not clear when this practice was introduced (Keown 2004, p. 297), it is valued as a way of affirming ones commitment to the Buddha's teachings and acquiring and transmitting merit by temporarily living in accordance with the dhamma. The stay might range from one day to several weeks, during which they have their heads shaved, wear monastic robes, reside in a nearby temple, and devote themselves to religious practices, including meditation and chanting. In recent years, there have been several cases of high-profile pilgrimage missions involving government employees from Thailand, including police and judges who resided at the Royal Wat Thai in Bodhgaya during the course of their ordination period.

Although the term "conversion" is somewhat problematic in the Indic context with respect to more fluid forms of South Asian religiosity and expression, it has become more prevalent in the modern era among people belonging to scheduled castes and scheduled tribes following the example of Bhim Rao Ambedkar. Much has been written about the highly accomplished low-caste leader Dr. Ambedkar (1891–1956) and his program of social reform that culminated in the highly symbolic conversion to Buddhism on 14 October 1956 in Nagpur, Maharashtra (Fitzgerald 1997; Jondhale and Beltz 2004; Zelliot 2008). Following in footsteps of Ambedkar, the majority of New Buddhists consider *diskha bhoomi* in Nagpur as a pilgrimage place and take initiations there. However, there are several instances of "conversion" ceremonies (or rather, *dhamma parivartan*) that have taken place under the boughs of the Bodhi tree or in other pilgrimage locales and sometimes coordinated though the Triratna Buddhist community (formerly TBMSG/FWBO) that organize these initiations among the Dalit community.

3.7. Consumption and Shopping

It is well known that pilgrimage and commerce are closely intertwined throughout the world. The record of Buddhist pilgrimage to India speaks to the importance of collecting sacred objects, "souvenirs", and mass reproductions that are believed to capture the spirit of a place, serve as a mnemonic device for storytelling, and imbue the pilgrim with the power of the original place visited. This is certainly the case for the reproduction of Mahabodhi temple miniatures (as well as large-scale reproductions) that commemorate the Buddha's enlightenment and have been found in several collections around the world (Asher 2012; Guy 1991). In more recent times, we have seen a growing market for various technologies of encapsulating the essence or authenticity of the pilgrimage journey that can be brought back to one's home and circulated among friends and relatives as gifts. Although postcards and prints have lost some of their appeal in recent years because of digital cameras, they continue to play an important role in documenting forms of devotion that help to capture a "sense of place" and as a way of memorializing the journey. Moving beyond the flatness of texts and pictures, a number of sacred souvenirs bring a different sense of materiality and are popular purchases in these sacred locales. Those in wide circulation include bodhi leaves, sandalwood perfume, wooden and stone buddha statues and images, *mala*-beads, Buddha lockets, meditation bags, incense sticks, singing bowls, bracelets, and thangka paintings (Figure 9).

Figure 9. Sacred souvenirs on display in the bazaar, Bodhgaya. (Photo from the author).

Some destinations have also gained certain notoriety for specific purchases that speak to the historical and mythological context for their sacredness such as in Rajgir/Nalanda—black Buddha image; Vaishali—wooden Ashoka pillar with the Lion on top; Sarnath—the iconic Buddha image housed in Sarnath Museum; Kuśinagarī—reclining Buddha image; and Lumbini—the baby standing Buddha image. In addition to these "sacred souvenirs", there is also a growing market among some pilgrim groups for certain Indian products such as silk scarfs and shawls, medicinal items, and Himalayan products, such as tea, herbal medicines, and popular Indian cosmetic brands, such as skin and face creams.[8]

4. Discussion: Embodying the Buddha-Dharma

Although we often interpret and imagine Buddhism as a religion of transcendence, in this concluding section, we draw insights from actor-network theory and material religion that takes seriously the close ontological ties between ritual practice and religious objects in mediating the human–divine dyad and destabilizing certain dichotomies such as the sacred/profane and spirit/matter. A question that often arises in the context of Buddhist pilgrimage, why is there a persistence of place in a religion that emphasizes *annica* or impermanence as part of the ethical foundations of Buddhist life and practice?

If *annica* is one of the three marks of existence in Buddhism—the others being *dukka* (unsatisfactoriness) and *anatta* (non-selfhood)—the term expresses the notion that all conditioned existence, without exception, is in a constant state of flux—a complete absence of permanence and continuity. This is exemplified in human life itself with the aging process and the subsequent cycle of birth and death (*samsara*). Thus, underlying the doctrine of *annica* are certain ontological assumptions regarding the timeless and limitless truth about reality and the physical embodiment of that truth through individual Buddhas and their material or corporeal remains.

As a means of addressing the continued presence and agency of the Buddha following his passing, we have shown how pilgrimage and the rituals embodied at sacred places play an integral role in the memory work that help to reinforce the patterns of religious life. On one level, sacred places and their material affordances provide important physical links to religious narratives that act as mnemonic devices for recalling sacred events in the biographic life of the founder while providing a paradigmatic model of aspiration for those who have made progress along the path to nirvana. These aspirations continue to play a formative role in linking the human world with transcendent realms as the moral axis points for the religious community. Thus, as places of memory, the Buddhist pilgrimage sites provide a ritual environment that helps to reinforce and internalize the truths of the dhamma because they help bring to life the stories of the Buddha in the holy land in ways that are morally transformative. As Cook (2018, p. 51) writes, "These practices are intended to effect changes within themselves, resulting in the actualization of moral good in accordance with Buddhist cosmology".

On another level, sacred sites are also invested with the presence of the Buddha by virtue of various techniques of ritualized consecration and worship involving sacred objects. The most widespread ritual technique for defining sacred place is relic enshrinement. For Buddhist adherents, it is believed that relics contain the energy of the Buddha and contact or proximity to these sacred objects yields strong soteriological benefits as sources of great merit (Huber 2008; Trainor 2007, p. 637). Through the symbolic construction of stupas and reliquary shrines, these material embodiments also suffuse and permeate the surrounding environment with the sacred and moral qualities of enlightened beings (Huber 2008, p. 61). Furthermore, as Huber points out with reference to Tibetan pilgrims, the superior quality of a holy place and its affective residue can also "be harvested and carried away, such as is commonly seen in the collection of water, earth and stones, or talismans from holy places by pilgrims" (2008, p. 61).

Thus, at a phenomenological level, the various ritual practices such as prostrations, chanting, and circumambulations not only provide a "focusing lens" (Smith 1980) that helps to evoke various affective responses and emotions such as *samvega*, but they also help to purify those within it. According to Anne Klein (Lama Rigzin Drolma), this is the "operative premise" of Buddhist ritual praxis. Not only do formal ritual elements and their repetition provide a great source of support along the path, but they also play an integral role in dissolving certain negative impediments toward the sublime goal of enlightenment. As Klein (2016) explains in the context of Tibetan ritual practice, this quality of "dissolving and evolving" is a vital key to the embodied experience of ritual that corresponds with the basic meaning of enlightenment: "the first is *byang*, meaning to cleanse, eliminate, remove; hence, dissolve. The second is *chub*, meaning to grow, blossom, further, develop; hence, evolve".

In the context of postcolonial modernity, Buddhist sacred space brings to the foreground a complex set of relationships and influences that converge around the notion of heritage and monumentality. There has been considerable scholarship in recent years around the ways in which heritage regimes (Geismar 2015) operate through an objectified historicist lens in ways that refashion relations among people giving rise to a sense of "disenchantment" and "social abstraction" that masks the exploitative and capital-intensive projects launched by various state governments and interest groups (del Marmol et al. 2013). As Michael Herzfeld describes, "Monumentality implies permanence, eternity, the disappearance of temporality except in some mythological sense. Monuments have a metonymic relationship to the entities (such as nation-states) that they serve, and their ponderous ontology discourages thoughts of their potential impermanence ... " (Herzfeld 2006, p. 129). These projects and prescriptions of monumentality are often ushered through by state authorities and international experts in the name of conservation, protection, and tourism development that may run in opposition to the religious sensibilities and ritual praxis of sacred space by devotees according to their own traditions of worship. Many Buddhist sites are situated in archaeological landscapes and because of the fragile nature of these landscapes, the governing authorities may prohibit or curtail the performance of certain rituals. Due to these restrictions and limited opportunities to perform certain rituals, the arena of performance often shifts to nearby monasteries and their temples. For instance, in Sarnath, pilgrims want to touch the Dhamek stupa but cannot do so because of fear of excessive wear and tear. In other words, the restrictions on certain ritual performances can create a spatial disconnect with access to certain relics and sacred objects, which is an essential part of pilgrimage (Shinde 2020).

At the same time, these forms of ritual practice also have the potential to subvert the strategies of place-making by heritage regimes and provide glimpses of alternatives to reigning territorial notions of property and enclosure (Collins 2011, p. 129) through their ongoing dialogues with devotion. As Collins writes, "There is little doubt that a naive empiricism focused on a material culture conceived of as a collection of objects fails to recognize the diversity of ways that people employ objects in social action (2011, pp. 127–28)". It is for these reasons, in this paper, we have focused on the phenomenological ritual relationship with sacred space—the sensory, performative, storied, and embodied dimensions of place. Directly associated with the ritual praxis is the Buddhist belief that pilgrimage places have transformative power and that sacred power is contingent on the ongoing ritual reconsecration and renovation of the built environment.

Although efforts are being made to control the ritual activity through various heritage and conservation management instruments, we argue that this ritual relationship should be paramount in decision making because these embodied practices of remembrance and veneration underpin the religious values and meanings associated with these places. This emphasis on ritual behavior and the role of "practice" among Buddhist pilgrims also helps draw attention to the ways in which meaning and action interact in affectively charged ways to inculcate and reinforce cultural knowledge and religious-cosmological structures. As Bourdieu has shown, practice activates meaning, in that "space can have no meaning apart from practice" (Bourdieu 1977, p. 214). This presents unique management challenges for religious site managers and caretakers as they negotiate and meditate the fault lines of the physical and intangible qualities of sacred space and how certain objects, bodies, actions, and ideas constitute a ritual ecosystem. Despite the technocratic nomenclature around "consuming" places of heritage and worship, Buddhist adherents do not undertake pilgrimage to "consume" the built environment but are active participants in the restitution and reinvention of a living sacred site that has implications for management, including new global conservation mandates around UNESCO.[9]

Due to the sheer number of pilgrims from abroad, it is difficult to generalize the significance of the pilgrimage journey, given the different backgrounds, motivations, and experiences. Although this paper provides a broad descriptive overview of the ritual ecology of sacred sites along the pilgrims path in North India and Nepal, more case studies are needed among different national groups and lineages to shed light on how Buddhist themselves imagine Buddhism on a translocal and transhistorical scale and the ongoing significance that rituals have for the importance of pilgrimage in the Indo-Gangetic region.

Author Contributions: Both authors contributed equally to this work. Both authors wrote, reviewed, and commented on the manuscript. Both authors have read and agreed to the published version of the manuscript.

Funding: There is no source of funding for this paper.

Institutional Review Board Statement: Not applicable.

Informed Consent Statement: Not applicable.

Data Availability Statement: Not applicable.

Conflicts of Interest: The authors declare no conflict of interest.

Notes

1 It has been argued elsewhere that the story of Aśokāvadāna and his redistribution of relics into 84,000 parts became a kind of ideal model for pilgrimage that entailed a process of mythologizing the pilgrim—and the 'pilgrim builder extraordinaire' (Coleman and Elsner 1995, p. 174).

2 As Strong (2014, p. 55) illustrates, given the different position and meanings associated with each of the four sites, MPNS reminds us that not all places of pilgrimage would have had the same affective quality and emotional charge, and therefore, it is not particularly clear from the Buddha's discourse "what aspects of the Buddha were thought to move pilgrims at different sites". For example, were sites such as Bodhgaya valued more given its association with the Buddha's awakening in comparison to a place such was Lumbini, which would have been associated more with the merits stemming from the Buddha's previous lives?

3 An important distinction that is often made when referring to Buddhist relics is 'primary relics' or 'corporeal relics' associated with enlightened persons and secondary relics or 'relics of contact' that were previously in the possession of sacred persons or were used and touched by them (see Deeg 2014, p. 13).

4 Among some Buddhist communities, such as the Tibetan-Himalayan groups, Vesak is referred to as Sawa Dawa, and it is believed that the merit associated with one's wholesome actions and deeds are particularly auspicious and increases 100,000 times. In the East Asian calendar Shakyamuni's enlightenment and passage into *parinirvāṇa* occur in the twelfth and eleventh lunar months. For example, among the Japanese, this includes *nehan-e (parinirvāṇa* day) on 15 February, *hanamatsuri* on 8 April (Buddha's birthday), and 8 December as *shaka-Jodo-e* (Bodhi day or date of Enlightenment).

5 For more on Tibetan practices of circumambulation, including Bon practitioners who circumambulate counterclockwise, see https://www.lamayeshe.com/article/kyabje-zopa-rinpoches-advice-circumambulation (accessed on 13 February 2021) and https://www.yowangdu.com/tibetan-buddhism/circumambulation.html (accessed on 13 February 2021).

6 Among some Tibetan-Himalayan Buddhists, prostrations are also frequently used in tandem with forms of visualization to a tantric deity. These prostrations may include the visualization of a particular "refuge tree" of which Guru Rinpoche (a common epithet of Padmasambhava) may be the centerpiece among the Nyingma sect, while reciting a prayer of taking refuge or the seven-line supplication to Guru Rinpoche. We would like to acknowledge Arthur McKeown for helping to clarify this.

7 Although the explanations for this exact number vary widely, some scholars attribute the number to a list of 108 afflictions, while others associate it with Buddhist cosmological renderings of all phenomenal existences (Buswell and Lopez 2013, p. 380)

8 Thank you to Manish Kumar for his insights on this section related to these items of purchases.

9 One expression of this is the Kyiv Statement on the protection of religious properties within the framework of the World Heritage Convention that emerged in 2010 following an international seminar on the role of religious communities in the management of World Heritage properties, involving active participation by religious authorities.

References

Asher, Frederick M. 2012. Bodh Gaya and the issue of originality in art. *Cross-Disciplinary Perspectives on a Contested Buddhist Site: Bodh Gaya Jataka* 7: 61.

Askew, Marc. 2007. The symbolic economy of religious monuments and tourist-pilgrimage in contemporary Thailand. In *Religious Commodifications in Asia: Marketing Gods*. Edited by Pattana Kitiarsa. New York: Routledge, p. 89.

Bourdieu, Pierre. 1977. *Outline of the Theory of Practice*. Translated by Richard Nice. Cambridge: Cambridge University Press.

Bruntz, Courtney, and Brooke Schedneck. 2020. *Buddhist Tourism in Asia*. Honolulu: University of Hawaii Press.

Buswell, Robert E., Jr., and Donald S. Lopez Jr. 2013. *The Princeton Dictionary of Buddhism*. Princeton: Princeton University Press.

Castelli, Elizabeth. 2003. "Editor's Preface". In *Connected Places: Region, Pilgrimage and Geographical Imagination in India*. Edited by Feldhaus Anne. New York: Palgrave MacMillan.

Choe, Jae Yeon, and Michael O'Regan. 2015. Religious Tourism Experiences in South East Asia. In *Religious Tourism and Pilgrimage Management: An International Perspective*. Edited by Razzaq Raj and Kevin Griffin. Oxfordshire: CABI International.

Coleman, Simon, and John Elsner. 1995. *Pilgrimage: Past and Present in the World Religions*. Cambridge: Harvard University Press.

Collins, John F. 2011. Culture, Content, and the Enclosure of Human Being: UNESCO's "Intangible" Heritage in the New Millennium. *Radical History Review* 2011: 121–35. [CrossRef]

Cook, Joanna. 2018. Remaking Thai Buddhism through International Pilgrimage to South Asia. In *Religious Journeys in India: Pilgrims, Tourists, and Travelers*. Edited by Andrea Marion Pinkney and John Whalen-Bridge. Albany: SUNY Press, pp. 37–64.

Deeg, Max. 2014. Buddhist Pilgrimage—An Introduction. In *Searching for the Dharma, Finding Salvation—Buddhist Pilgrimage in Time and Space*. Edited by Christoph Cueppers and Max Deeg. Lumbini: Lumbini International Research Institute, pp. 1–21.

del Marmol, Camila, Marc Morell, and Jasper Chalcraft. 2013. *The Making of Heritage: Seduction and Disenchantment*. New York: Routledge, vol. 8.

Eck, Diana L. 2012. *India: A Sacred Geography*. New York: Harmony.

Fitzgerald, Timothy. 1997. Ambedkar Buddhism in Maharashtra. *Contributions to Indian Sociology* 31: 225–51. [CrossRef]

Geary, David, and Sraman Mukherjee. 2017. Buddhism in Contemporary India. In *The Oxford Handbook of Contemporary Buddhism*. Edited by Michael Jerryson. Oxford: Oxford University Press, pp. 36–60.

Geary, David. 2008. Destination enlightenment: Branding Buddhism and spiritual tourism in Bodhgaya, Bihar. *Anthropology Today* 24: 11–14. [CrossRef]

Geary, David. 2014. Rebuilding the Navel of the Earth: Buddhist pilgrimage and transnational religious networks. *Modern Asian Studies* 48: 645–92. [CrossRef]

Geary, David. 2017. *The Rebirth of Bodh Gaya: Buddhism and the Making of a World Heritage Site*. Seattle: University of Washington Press.

Geismar, Haidy. 2015. Anthropology and heritage regimes. *Annual Review of Anthropology* 44: 71–85. [CrossRef]

Gibson, James J. 1977. The theory of affordances. In *Perceiving, Acting, and Knowing, Toward an Ecological Psychology*. Edited by Robert Shaw and John Bransford. New York: Halsted Press Division, Wiley, pp. 67–82.

Goldberg, Kory. 2013. Pilgrimage Re-oriented: Buddhist Discipline, Virtue and Engagement in Bodhgayā. *The Eastern Buddhist* 44: 95–120.

Guy, John. 1991. The Mahābodhi temple: Pilgrim souvenirs of Buddhist India. *The Burlington Magazine* 356–67.

Hall, C. Michael. 2006. Buddhism, tourism and the middle way. In *Tourism, Religion and Spiritual Journeys*. Edited by Dallen J. Timothy and Daniel H. Olsen. New York: Routledge, pp. 172–85.

Herzfeld, Michael. 2006. Spatial cleansing: Monumental vacuity and the idea of the West. *Journal of Material Culture* 11: 127–49. [CrossRef]

Huber, Toni. 2008. *The Holy Land Reborn: Pilgrimage and the Tibetan Reinvention of Buddhist India*. Chicago: University of Chicago Press.

Hung, Kam, Xiaotao Yang, Philipp Wassler, Dan Wang, Pearl Lin, and Zhaoping Liu. 2017. Contesting the commercialization and sanctity of religious tourism in the Shaolin Monastery, China. *International Journal of Tourism Research* 19: 145–59. [CrossRef]

Hung, Kam. 2015. Experiencing buddhism in chinese hotels: Toward the construction of a religious lodging experience. *Journal of Travel & Tourism Marketing* 32: 1081–98.

Jondhale, Surendra, and Johannes Beltz. 2004. *Reconstructing the World: BR Ambedkar and Buddhism in India*. Oxford: Oxford University Press.

Keown, Damien. 2004. *A Dictionary of Buddhism*. Oxford: OUP Oxford.

Kerestetzi, Katerina. 2018. The spirit of a place: Materiality, spatiality, and feeling in Afro-American religions. *Journal de la société des américanistes* 104. [CrossRef]

Kitiarsa, Pattana. 2010. Missionary intent and monastic networks: Thai Buddhism as a transnational religion. *Sojourn: Journal of Social Issues in Southeast Asia* 25: 109–32. [CrossRef]

Klein, Anne C. 2016. Revisiting Ritual. *Tricycle*. Available online: https://tricycle.org/magazine/revisiting-ritual/ (accessed on 13 February 2021).

Maitland, Padma D. 2018. The Object/The Tree: Emissaries of Buddhist Ground. In *Records, Recoveries, Remnants and Inter-Asian Interconnections*. Edited by Anjana Sharma. Singapore: ISEAS Publishing, pp. 160–83.

Mason, David, and Moo-Hyung Chung. 2018. The Burgeoning of the Baekdu-daegan Trail into a New Religious-Pilgrimage Tourism Asset of South Korea. *Journal of Tourism and Leisure Research* 20: 425–41.

Maud, Jovan. 2017. Buddhist relics and pilgrimage. In *The Oxford Handbook of Contemporary Buddhism*. Edited by Michael Jerryson. Oxford: Oxford University Press, pp. 421–35.
McDaniel, Justin Thomas. 2016. *Architects of Buddhist Leisure*. Honolulu: University of Hawai'i/Hawai'i Press.
Mills, Laurence-Khantipalo. 1982. *Lay Buddhist Practice: The Shrine Room, Uposatha Day, Rains Residence*. Kandy: Buddhist Publication Society.
Mukherjee, Sraman. 2018. Relics in Transition: Material Mediations in Changing Worlds. *Ars Orientalis* 48: 20–42. [CrossRef]
Nugteren, Albertina. 1995. Rituals around the bodhi-tree in Bodhgaya, India. In *Pluralism and Identity*. Edited by J. Platvoet and K. van der Toorn. Leiden: Brill, pp. 143–65.
Ober, Douglas F. 2019. From Buddha Bones to Bo Trees: Nehruvian India, Buddhism, and the poetics of power, 1947–1956. *Modern Asian Studies* 53: 1312–50. [CrossRef]
Philp, Janette, and David Mercer. 1999. Commodification of Buddhism in contemporary Burma. *Annals of Tourism Research* 26: 21–54. [CrossRef]
Preston, James J. 1992. Spiritual Magnetism: An Organising Principle for the Study of Pilgrimage. In *Sacred Journeys: The Anthropology of Pilgrimage*. Edited by Alan Morinis. Westport: Greenwood Press, pp. 31–46.
Queen, Christopher S. 2012. *Engaged Buddhism in the West*. New York: Simon and Schuster.
Ray, Himanshu Prabha. 2014. *The Return of the Buddha: Ancient Symbols for a New Nation*. New York: Routledge.
Schopen, Gregory. 2004. *Buddhist Monks and Business Matters: Still more Papers on Monastic Buddhism in India*. Honolulu: University of Hawaii Press, vol. 2.
Seelananda, Ven Bhikkhu T. 2010. *Footprints of the Buddha (Pilgrimage to Buddhist India)*. Edmonton: Samatha-Vipassana Meditation Centre.
Sharf, Robert H. 1999. On the allure of Buddhist relics. *Representations* 66: 75–99. [CrossRef]
Sharma, Anjana. 2018. *Records, Recoveries, Remnants and Inter-Asian Interconnections: Decoding Cultural Heritage*. Singapore: ISEAS-Yusof Ishak Institute.
Shinde, Kiran. 2020. The spatial practice of religious tourism in India: A destinations perspective. *Tourism Geographies*, 1–21. [CrossRef]
Singh, Upinder. 2010. Exile and return: The reinvention of buddhism and buddhist sites in modern India. *South Asian Studies* 26: 193–217. [CrossRef]
Smith, Jonathan Z. 1980. The bare facts of ritual. *History of Religions* 20: 112–27. [CrossRef]
Strong, John. 2014. The beginnings of Buddhist pilgrimage: The four famous sites in India. In *In Searching for the Dharma, Finding Salvation—Buddhist Pilgrimage in Time and Space*. Edited by Christoph Cueppers and Max Deeg. Lumbini: Lumbini International Research Institute, pp. 49–63.
Trainor, Kevin. 1997. *Relics, Ritual, and Representation in Buddhism: Rematerializing the Sri Lankan Theravada Tradition*. Cambridge: Cambridge University Press, vol. 10.
Trainor, Kevin. 2007. Sacred Places. In *Encyclopedia of Buddhism*. Edited by Damien Keown and Charles S Prebish. New York: Routledge, pp. 633–42.
Wallace, B. Alan. 2002. The spectrum of Buddhist practice in the West. *Westward Dharma: Buddhism Beyond Asia*, 34–50.
Wong, Cora Un In, Alison McIntosh, and Chris Ryan. 2013. Buddhism and tourism: Perceptions of the monastic community at Pu-Tuo-Shan, China. *Annals of Tourism Research* 40: 213–34. [CrossRef]
Wong, Cora Un In, Alison McIntosh, and Chris Ryan. 2016. Visitor management at a Buddhist sacred site. *Journal of Travel Research* 55: 675–87. [CrossRef]
Zelliot, Eleanor. 2008. Understanding Dr. BR Ambedkar. *Religion Compass* 2: 804–18. [CrossRef]

Article

Contested Authenticity Anthropological Perspectives of Pilgrimage Tourism on Mount Athos

Michelangelo Paganopoulos

Global Inquiries and Social Theory Research Group, Faculty of Social Sciences and Humanities, Ton Duc Thang University, Ho Chi Minh City, Vietnam; michelangelopaganopoulos@tdtu.edu.vn

Abstract: This paper investigates the evolution of customer service in the pilgrimage tourist industry, focusing on Mount Athos. In doing so, it empirically deconstructs the dialectics of the synthesis of "authentic experience" between "pilgrims" and "tourists" via a set of internal and external reciprocal exchanges that take place between monks and visitors in two rival neighboring monasteries. The paper shows how the traditional value of hospitality is being reinvented and reappropriated according to the personalized needs of the market of faith. In this context, the paper shows how traditional monastic roles, such as those of the guest-master and the sacristan, have been reinvented, along with traditional practices such as that of confession, within the wider turn to relational subjectivity and interest in spirituality. Following this, the material illustrates how counter claims to "authenticity" emerge as an arena of reinvention and contestation out of the competition between rival groups of monks and their followers, arguing that pilgrimage on Athos requires from visitors their full commitment and active involvement in their role as "pilgrims". The claim to "authenticity" is a matter of identity and the means through which a visitor is transformed from a passive "tourist" to an active "pilgrim".

Keywords: authenticity; spirituality; Mount Athos; hospitality; pilgrimage tourism

Citation: Paganopoulos, Michelangelo. 2021. Contested Authenticity Anthropological Perspectives of Pilgrimage Tourism on Mount Athos. *Religions* 12: 229. https://doi.org/10.3390/rel12040229

Academic Editor: Kiran Shinde

Received: 14 February 2021
Accepted: 21 March 2021
Published: 25 March 2021

1. Introduction

In investigating the dialectics between the concepts of "tourism" and "pilgrimage landscape", and "tourists" and "pilgrims", respectively, this paper uses two key *claims* that affirm the "sacredness" of pilgrimage tourism: "authenticity" and "spirituality". In doing so, the paper focuses on two rival monastic communities situated on Mount Athos, an autonomous Christian Orthodox monastic republic of twenty self-sustainable monasteries and their settlements, situated in the mountainous peninsula of Chalkidiki in northern Aegean. In the classic terms of Edward Bruner (2005, p. 18), Mount Athos functions as a "touristic borderzone", a crossroad where the monastic and secular worlds meet and reciprocate. At the time of my fieldwork, the estimated number of visitors to Athos was up to 50,000 (Greek newspaper *Makedonia*, 28 November 2005, p. 31) and reportedly increased to more than 75,000 by 2018[1] (as there is an unknown number of informal visitors whose close relationship to monks overrides formal entry regulations). The paper focuses on the years between 2002 and 2004, which anticipated the rapid increase in the estimated number of visitors to the peninsula of Athos. The paper explores some of the economic solutions given to the challenges represented by the increasing presence of visitors in the daily life of the monasteries, including imposing and reinforcing rules and prohibitions associated with the land, a new timetable and divisions in space, and other "economic" (i.e., compromised) changes and reinvention of practices, such as confession. In turn, the changes in the way the monks relate to their guests affect traditional roles, such as those of the confessor evolving into that of the life-advisor, as well as monastic practices

1 "Elder Ephraim of Mt. Athos speaks with TNH" (10 February 2018). Available online: https://www.thenationalherald.com/archive_church/arthro/elder_ephraim_of_mt_athos_speaks_with_tnh-19466/ (accessed on 2 December 2020).

relating to these roles, such as confessions evolving into psychoanalytical sessions. By comparing two opposite *claims to authenticity* by the rival brotherhoods and their followers of the monasteries of Vatopaidi and Esfigmenou, the paper deconstructs the dialectics of Athonian "hospitality" (*"filoxenia"*) on the basis of a set of internal and external reciprocities taking place on the holy grounds of the monasteries between "the worlds of the laity and the celibate monasteries" (Loizos and Papataxiarchis 1991, p. 16).

2. Fieldwork

The material derives from my own personal experiences back in 2002 to 2004, following my journeys to two monasteries of Mount Athos.[2] By asking the permission of the monks, they kindly allowed me to stay for long periods in the monasteries, where I conducted fieldwork in a systematic and methodological way, gathering empirical data using interviews, archive and administrative material. I also extensively participated in the everyday life of the monasteries, while casually talking with visitors and monks. The quality of service and the things expected of me in each monastery revealed to me a very different understanding and ways of offering hospitality (*"filoxenia"*), one of the main traditional services of the monasteries. Respectively, the means and quality of hospitality I received in each monastery gave me a rare insight into the differences in customization and services offered to visitors. In the monastery of Vatopaidi, the monks agreed for me to stay and conduct my research on the basis that I would fully participate in the liturgical and working life of the monastery. For this reason, my personal "informants" became my "spiritual father" (*"pneumatikos pateras"*) who supervised me during my confessions to him in the night, and the guest-master on duty (*"archontaris"*) who supervised me at work during the day at the guesthouse. By contrast, in the monastery of Esfigmenou, I was not allowed to reciprocate my staying with manual work or confession. The monks of this monastery consider frequent confessions to be a "Western" influence and not as part of their "pure" zealot tradition. In similar terms, I was not allowed to help or contribute in any way for my staying. Rather, I was only meant to pray throughout my staying and take the oath to the "Old Calendarist Church", an ultra-Orthodox sect, as part of this monastery's interpretation of hospitality.

The paper qualitatively compares the customization of hospitality in the two monasteries because they represent two opposite environments whose ethnographic comparison highlights both the heterogeneity and openness in conceptualizing and reinventing the Athonian tradition in terms of its "authenticity". Furthermore, at the time of my fieldwork, major developments took place in Esfigmenou, as the monastery had been placed under embargo (and still remains so) due to its politics outside Athos, which meant that no visitors were allowed to approach its premises. However, many found alternative, and at times, illegal ways of entering the monastery and supporting the monks with their political cause. This raised a public controversy regarding the Esfigmenou's uses of hospitality and legitimacy of services offered by its monks to pilgrims in exchange for ideological loyalty from visitors.

A second reason for reflecting upon the material gathered during this period, i.e., the turn of the millennium, is the beginning of the flourishing of the tourist pilgrimage industry, following the global turn to subjectivity and "spirituality", especially amongst Christians (Speake and Conomos 2005; Speake and Gothóni 2008; and Heelas et al. 2005). The global turn to spirituality saw a sharp rise and development of "pilgrimage tourism" and an ever-increasing interest in religious sites as tourist attractions (Rejman et al. 2016, p. 568). A part of this wider process of "re-enchantment" with religion (Comaroff and Comaroff 2000, pp. 291–343) was the rise of religious tourism, emerging within a

[2] As part of my doctoral research, I visited Athos ten times over periods that lasted from four to eight weeks. I wish to kindly thank the monks for their hospitality, discretion, interest, and their general positive input to the research. Trip 1: 19 February 2002 Vatopaidi, Trip 2: 29 November 2002 Esfigmenou, Trip 3: 5 January 2003 Esfigmenou, Trip 4: 25 February 2003 Esfigmenou, Trip 5: 25 March 2003 Vatopaidi, Trip 6: 6 June 2003 Esfigmenou (jumping fence because of embargo), Trip 7: 2 August 2003 Vatopaidi, Trip 8: 12 September 2003 Vatopaidi, Trips 9–10: August 2004 Vatopaidi and Esfigmenou.

reinvented and personalized market of faith (as in Kim et al. 2020 and Wang et al. 2020). Religious tourism as a pilgrimage experience is a contemporary phenomenon that defies a specific definition due to the intermarriage between tourist and religious experiences (see Tilson 2005, pp. 9–40; Olsen and Timothy 2006, pp. 1–22, and Rejman et al. 2016, pp. 562–75). The rapid changes in customer service in monastic settings reflected upon the wider transformation of the monastic economies from "local economies" into global networks that emerged through the rising "capitalist global economy" (Iossifides 1991, pp. 136–37). They raised dilemmas over *how* to appropriate "authentic" traditional values against global processes of "de-traditionalization" (Heelas et al. 1996), as a means of preserving both the economic and ritual life of the monasteries (Tanasyuk and Avgerou 2009, online source).

The paper investigates the "authentic" monastic ideal invested in the value of hospitality on Athos, as it emerges out of the dialectics between "re-enchantment" and "de-traditionalization" played by the paradoxical presence of visitors in the daily life of the monasteries. In these terms, the comparative material reveals how the openness of sacred sites to the world via the pilgrimage industry raises moralized questions and tensions regarding *claims to "authenticity"*, especially amongst Christians who dogmatically juxtapose money to faith as a means of separating "tourists" from "pilgrims", respectively.

The paper aims to theoretically and empirically contribute to the development of tourism theory for pilgrimage sites, and the wider ongoing discussion over cultural and economic tensions and challenges between development, sustainability, and cultural preservation of monasteries and their authentic way of life. In this context, the paper views hospitality as the grey conceptual area where the monastic and secular worlds meet via a set of reciprocities and mutual agreements. By expanding into personal ("spiritual") and collective (ideological) motivations that attract a visitor to a *specific* monastery, the paper empirically illustrates: (a) how "objective" mechanisms *objectify and naturalize* specific cultural settings, by transforming *all* visitors into "pilgrims" through a cathartic "spiritual" system of practices such as confession; and (b) how, in the politicized context of Esfigmenou, the "pilgrims" then become activists supporting the ultra-Orthodox ideology of the "Authentic Christians", which deems the other monasteries as being less, or even, *not* being "authentic".

3. Authenticity and Pilgrimage Tourism

The millennial turn to subjectivity and "spirituality" saw the rise and development of "pilgrimage tourism" and an ever-increasing interest in religious sites as tourist attractions (Rejman et al. 2016, p. 568). The concept lacks a specific definition due to the ambiguity characterizing the tourist industry in respect to evaluating and integrating subjective motivations and personal experiences as part of the pilgrimage tourist package (see Tilson 2005, pp. 9–40; Olsen and Timothy 2006, pp. 1–22, and Rejman et al. 2016, pp. 562–75). As previous anthropological literature on the subject shows (Morinis 1992), pilgrimage is a process of both self-transformation and self-affirmation via collective participation. Furthermore, beyond the experience of sightseeing, the process formulates social relations made on the way to the site, for example, bonding or arguing between members of the same or a rival group. Ethnographies on Christian pilgrimage show that the journey to a sacred site may become an arena of contestation between rival groups of pilgrims (Eade and Sallnow 1991; Bax 1983, pp. 167–77; Bax 1990, pp. 63–75, and Sallnow 1981, pp. 163–82).[3] Victor Turner (1974, 1997) has extensively discussed the emotional identification with the creation of an identity through transformative rites of passages, such as pilgrimages, highlighting their social value in creating spontaneous cohesion and bonding between members of the same group (i.e., a "*communita*") via their journey to a pilgrimage site. The

[3] Eade and Sallnow (1991), Bax (1983, pp. 167–77, 1990, pp. 63–75), and Sallnow (1981, pp. 163–82) among others, previously challenged Victor and Edith Turner's Durkheimian concept of "*communitas*" (Turner 1997, 1974) by pointing to the competitive nature of pilgrimage over religious sites. Elsewhere, other examples from rural Greece equally show how such competitive traditional practices define the formation and contestation of collective gendered identities (as in Herzfeld 1985, and Seremetakis 1991) and more recently, matters of historical claims and identity (Solomon 2021).

journey initiates a process of self-transformation, which involves group experiences and shared memories emerging within the group on their way to a shrine.

Hence, to reduce pilgrimage to a sightseeing experience excludes personal and emotional motivations taking place *during* a pilgrimage, as well as any "spiritual" esoteric changes that may have an impact on the personality and lifestyle of the traveler in the long term: "rigid dichotomies between pilgrimage and tourism, or pilgrims and tourists, no longer seem tenable in the shifting world of postmodern travel" (Badone and Roseman 2004, p. 2). Indeed, in their overview regarding the ambiguity of terms such as "religious tourism", Kim et al. (2020, online source) concluded that, "Understandings of religious tourism have evolved beyond pilgrimage and now encompass the meaningfulness of a destination." These may include "individual religious affiliations" and a personal and subjective sense of identity and religiosity associated with the history and vibes of the place (2020, pp. 185–203). As in Bauman's classic analysis of "postmodern religion" (Bauman 1998, pp. 55–78), sacred places project a sense of a concrete identity—in an increasingly fragmented and confusing postmodern world. Accordingly, each monastic community packages a very different, and even antagonistic, "pilgrim experience" to its visitors in relation to the other monasteries, which demands their *active* involvement, by being fully engaged in the antagonistic nature of the neoliberal market of faith—sometimes even against each other.

Central in enhancing this sense of identity and belonging during a pilgrimage is the "claim to authenticity" of a place and/or a way of life (as in Theodossopoulos 2013, pp. 337–60), which implies the existence of multiple authenticities, or rather, *claims and degrees to authenticity*. I discuss this in terms of tourist satisfaction in relation to the alleged anticommercial service of the monks to visitors. Studies in tourism, discuss "authenticity" as a means of attraction and satisfaction. Zatori et al. (2018) and Zhang et al. (2018) both quantitatively investigated tourist satisfaction in terms of experience involvement and interactive processes of customization (Zatori et al. 2018, pp. 113–15, and Zhang et al. 2018, p. 927). Zatori et al. pointed to authenticity and memorability as central in offering tourist satisfaction along with the "experience involvement" the hosts allow to their visitors. The latter play an active, "important role in experience formation and value co-creation with the service provider" (Zatori et al. 2018, p. 111). The authors use MacCannell's classic model of "staged authenticity" which develops over three overlapping evaluations: (a) its objectified value; (b) constructed cultural determinants; and (c) in terms of existential authenticity. Wang (1999) and Zhang et al. (2018) associated each stage with a particular trend of thought, i.e., objectivity (focusing on the "objective" value of relics and the landscape); constructivism (the vocation of the place in tourist brochures and its reputation in the market); and postmodernism (offering a transformative experiential process of self-discovery as part of the services), respectively (MacCannell 1973, pp. 589–603, and in Wang 1999, pp. 349–70, and Zhang et al. 2018, pp. 927, among others). The first two components of "authenticity" (objective and constructive authenticity) relate to the place and its material objects. Experiential authenticity, on the other hand, focuses on the experiential impact and personal transformation a pilgrimage may have on a visitor in the long term.

In this context, Zatori et al. (2018) defined "customization" as a two-way co-creative process, which emerges in the interaction between visitors and tourist operators during guided tours according to their needs and expectations. In their research on interactive customization, Zatori et al. highlighted the importance involvement plays in forming a positive tourist experience: "The extent to which an experience can engage an individual and how important the experience becomes depends on the extent of the interaction between the company and the customer." (Zatori et al. 2018, pp. 112–13). The interaction and commitment of a pilgrim to a certain setting and/or bonding to a monk can extend to joining their hosts in their ideological battles. "Real-time experiences" (ibid. p. 114) thus co-create the path of self-discovery and self-transformation, which, in turn, enhances the masculine bond between the visitor and the monk in the role of a tourist operator, on the basis of their shared and staged "authentic experience". However, since experience and memorability cannot be fully quantitatively evaluated, the existential and influential aspects of authenticity represent a methodological difficulty for the evaluation of pilgrimage tourism in depth (see also Zatori et al. 2018, pp. 114–18; Tilson 2005, pp. 9–40; Olsen and Timothy 2006, pp. 1–22, and Rejman et al. 2016, pp. 562–75).

A second issue associated with the specificity of the landscape is the "anti-commercial" and/or "non-commercial" attitude of the monks of Athos towards the tourist industry, due to the free nature of the "spiritual" services the monks offer to visitors Speake (2002) paints a nostalgic and ideal picture of a kind of premodern experience of living, as if it was extracted from the Byzantine Roman past. Athos offers an "alternative to the rapidly materialism and secularism of modern society" (Speake 2002). On the other hand, Andriotis argues that Athos "has not been developed with the intention of providing a landscape for 'sale' or 'consumption'" (Andriotis 2011, p. 1624). Andriotis then associates the "non-commercial nature" of Athonian Hospitality with the collective *feeling* of authenticity associated with the sacred landscape., The comparative material of this paper shows that there is no "objective" or "constructive" value given to the Athonian sense of "authenticity". As Cohen (1988, p. 371) exclaimed, "authenticity is conceived as a negotiable rather than primitive concept," which in the context of commoditization in tourism may evolve into various "emergent" authenticities. The way Athonian "authenticity" is marketed in tourist brochures and the web, projects a rather idealist way of reimagining the sacred peninsula in the present time as a means of attraction and tourist satisfaction (Selwyn 1996). The comparison of the interpretation and means of hospitality from Vatopaidi and Esfigmenou show that the vibe of "authenticity" is not a static, homogeneous, or singular tour-de-force, but rather, evolves into emergent authenticities, which may even contradict each other by questioning the "authenticity" of other counterclaims. This contestation then raises dilemmas over what is "authentic" and what may be a "merely spectacle" (as in Kim and Jamal 2007, pp 181–201).

Accordingly, the paper examines how each monastic community's claim to "authenticity" constitutes an *objectified collective feeling* (than a historical fact), socially emerging out of the dialectics of everyday life. In this context, the *claim to authenticity* may become a vehicle of a certain ideology that constitutes the vocation of each monastic institution outside Athos. The mutual existence of a number of authenticities further reveals the heterogeneity found on Athos in terms of customer categorization, offered services, and means of attraction, all of which depending on the dialectics between each *particular* monastic institution and its material history (objective and constructive authenticity), and the personalized services its monks offer to their visitors (existential and influential authenticity). The dialectics that determine this process of objectifying the value of "authenticity" are: [*THESIS*] Objective/Constructive authenticity (PLACE) vs. [*ANTITHESIS*] existential/influential authenticities (PERSON) => [EMERGING *SYNTHESIS*] Objectified Value of a pilgrimage landscape.

4. Athos: An "Authentic" Landscape and Way of Life

In my discussions with monks during my fieldwork, they considered themselves to be life-long visitors to the "Garden of the Mother of God" (*"o kipos tis Theotokou"*) or the "Garden of Virgin Mary" (*"o Kipos tis Parthenou"*). They believe that Mary is the owner (*"ktitorissa"*) of the peninsula, whose omnipotent presence traditionally establishes a "spiritual" way of monastic life (*"pneumatiki zoe"*) associated with the Holy Mount, as opposed to the "materialist" way of secular life associated with the "worldly world" (*"kosmikos kosmos"*) outside Athos. This separation is traditionally affirmed by the rule of *Avaton* ("No Trespassing") referring to the prohibition of all females from the peninsula, which dates back to the very foundation of the republic in the 9th century (Papachrysanthou 1992; Paganopoulos 2007, pp. 122–33, 2020, pp. 66–87). In this context, the peninsula's "sacredness" demonstrates the Durkheimian understanding of monastic life as a "sacred" way of life (Durkheim [1912] 1995, pp. 37–42; Ross 1991, p. 100, and Paganopoulos 2007, pp. 123–24, 2020, pp. 66–68). The exceptional "spirituality" associated with the peninsula is legally affirmed under the "spiritual" protection of the "Ecumenical Patriarchate" (as per paragraph 1 of Article 105 of the Greek Constitution, see also Hellenic Parliament website).[4] In 1979, the European Community further ratified the autonomous status of the republic, "justified exclusively on grounds of a spiritual and religious nature".[5] By 1992, the republic was further included in and funded by the UNESCO World Heritage Program for the preservation of its "authentic" way of life.[6]

UNESCO's inclusion of Athos as a World Heritage Site enhanced the peninsula's worldwide reputation for offering a personalized "authentic" experience of monastic life to its visitors, described in online blogs as "mystical", "spiritual", and "emotional", "[...] difficult to explain unless you have experienced that once in your life" (blogger Christian Zocca, 20 December 2020).[7] In breaking down this claim to "authenticity" into its components, Andriotis (2011, pp. 1613–33) made the correlation between "genres of authenticity" of pilgrimage landscapes (Gilmore and Joseph 2007), and the core elements of experiencing the sacred Athonian site (Andriotis 2009, pp. 64–84). For Andriotis, a pilgrimage landscape is an "experiential cultural space" (2011, p. 1614) consisting of five corresponding elements: natural/spiritual, original/cultural, exceptional environmental/referential/secular, and influential/educational, respectively. In deconstructing the "authentic" experience of visiting Athos, Andriotis highlighted a dialectical tension between supply and demand in terms of "objective authenticity" and "existential experience", respectively (Ibid). In this dialectical context, the "objective authenticity" of a place or an object depends on a Durkheimian collective recognition of "authenticity" as a *social phenomenon*: "[...] the authentic experience is caused by the recognition of the toured objects as "authentic" [...] not because they are inherently authentic but because they are constructed as such in terms of point of view, beliefs perspectives or powers" (Wang 1999, p. 351 cited in Andriotis Ibid., and also in Belhassen et al. 2008, p. 686).

Most of the visitors that I engaged with during my fieldwork, tended to highlight that they had travelled to Athos to get a taste of the monasteries' way of life as "it was a thousand years ago". This shared motivation revealed a collective nostalgia for a golden Byzantine Empire, or "structural nostalgia" in Herzfeld's terms of a "longing for an age before the state" (Herzfeld 1997, p. 22): an imagined and protected masculine space (which traditionally excludes the presence of women), where only male visitors can nostalgically and religiously experience an authentic Byzantine living past, by engaging with monks in their activities (as also in Andriotis' example of Athonian "original authenticity", 2011, p. 1622). Furthermore, the pan-Orthodox and international vocation of Athos to Christians from across the globe responds to ideals of a "Byzantine universalism" beyond the boundaries of modern nation states (as in Tzanelli 2008, pp. 141–50). Objects of veneration such as miraculous relics and ancient icons kept in monasteries (as in Andriotis' example of "referential authenticity" 2011, pp. 1625–26) attract thousands of visitors every year from across the Orthodox spectrum and beyond, supporting an informal industry of "pilgrimage tourism" that has globally emerged with an ever-increasing interest in visiting religious sites as tourist attractions (Rejman et al. 2016, p. 568). Vice versa, this nostalgic return to a reinvented Imperial past, associated with the sacredness of the landscape, enhances a

4 Available at: http://home.lu.lv/~rbalodis/Konst%20tiesibas/Federalisms/Feder_Unit%C4%81r%C4%81_Grie%C4%B7ija_Konstitucija.pdf; https://www.hellenicparliament.gr/en/Vouli-ton-Ellinon/To-Politevma/Syntagmatiki-Istoria/ (accessed on 17 January 2019).

5 Source: UNESCO/CLT/WHC (Official Journal L 291, 19/11/1979 P. 0186). Available online: http://whc.unesco.org/en/list/454.11979H/AFI/DCL/04/ (accessed on 17 January 2019).

6 UNESCO's "World Heritage List" describes Athos in terms of its "**Authenticity:** "The property reflects adequately the cultural values recognized in the inscription criteria through the setting of the monasteries and their dependencies, together with the form, design and materials of the buildings and farms, their use and function, and the spirit and feeling of the place. Mount Athos has an enormous wealth of historic, artistic and cultural elements preserved by a monastic community that has existed for the last twelve centuries and constitutes a living record of human activities". Available online: https://whc.unesco.org/en/list/454/, and also *see* "World Heritage Data Sheet" at https://yichuans.github.io/datasheet/output/site/mount-athos/ (accessed on 2 December 2020).

7 "Have a Mystical Experience in Halkidiki Mount Athos" (20 December 2020) in Zocca, Christian's blog "Trips Are Over." Available online: https://tripsareover.com/en/greece/experiences/have-a-mystical-experience-in-halkidiki-on-mount-athos/ (accessed on 2 February 2021).

deeper collective feeling of *authenticity* springing out of the separation of monastic from secular life outside the border of the republic.

For a visitor, this "'passage from the profane to the sacred' [. . .] serves as a journey to the world of Byzantium as well as an initial signifier of natural authenticity" in Andriotis' words (2011, p. 1620). Everyday life inside the monasteries follows the *coenobitic* (communal) rule introduced by St Athanasius the Athonite (920-1003AC) a millennium ago.[8] The peninsula lacks infrastructure, offering limited access to the monasteries. Although some monasteries use photovoltaic energy and generators for electricity, the majority of smaller settlements have no electricity, using instead oil lamps and stoves for heating. There are no asphalted roads or electric wires connecting them, but only paths crossing the thick forest leading to the southern cliffs of the peninsula. In their accounts, the monks live a "symbiotic" way of living within nature, which is directly associated with the sacredness of the landscape (Paganopoulos 2020, pp. 66–67). The monastic *coenobitic* program, timetable and activities follow the change of winter and summer solstices and natural succession of night and day. Since the nights are longer in the winter, the monks spend most of their time in their cells praying and "spiritually" cultivating their monastic self in privacy. In the summer, when the days are longer and warmer, they place more emphasis on daily activities and labor in the fields and the churches. This "symbiosis", therefore, combines traditional practices of faith with agricultural activities, naturalizing the presence of the monks and the monasteries as "natural" parts of the sacred landscape (as also in Andriotis' example of "natural authenticity", 2011, p. 1621).

The collective feeling of "authenticity" is central in the vocation of Athos in the Christian world. The thirst to taste the "authentic" life annually attracts thousands of visitors to Athos, who contribute both to the economy of the monasteries and to the entire local area situated in the touristic peninsula of Chalkidiki, north Greece. In this respect, the traditional prohibition of all females from Athos has been "economically" compromised by the development of an alternative sightseeing industry for "visiting" the monasteries. For example, private cruise companies offer to both men *and* women a "short pilgrimage and cultural break" as a "spiritual sailing experience"[9] around the peninsula for sightseeing without having to step on ground but remaining in the boat ("floating tourism", Kotsi 1999, pp. 5–20; Kotsi 2012, p. 154). Another example of reinvention of tradition by a younger generation of monks was the development of the traditional symbiotic relationship to the landscape towards a focus on ecological sustainability. The international group "Friends of Mount Athos" (FOMA) established in the 1990s an annual "spiritual ecology camp" taking place in the monasteries of Vatopaidi and Simonopetra. It consists of "a 12-day programme of activities that combine care of the environment on Mount Athos with the spiritual benefits gained from pilgrimage there and participation in the life and prayer of the monasteries".[10]

5. Accommodation in Vatopaidi

In general terms, the customer service the monasteries offer to visitors is an outcome of their strict internal hierarchical organization. In my very first visit to the monastery of Vatopaidi, for example, back in September 2002, it became evident to me that it is a

[8] St. Basil's *coenobitic* (communal) mode of life refers to a communal type of monastic living in which its members share spiritual and working labor, food and clothing, hierarchy and egalitarianism, rules and timetable, practices of faith and common spaces (Malavakis 1999 and Vergotis 1995, p. 126).

[9] Advert by "All in Bluesive Private Cruises." "Mount Athos: Special, Spiritual Sailing Experiences." Available online: https://allinblusive.co.uk/blog/mount-athos-special-spiritual-sailing-experience/ (accessed on 2 February 2021).

[10] Available online: https://athosfriends.org/visiting/join-the-syndesmos/ and https://athosfriends.org/events/2020-spiritual-ecology-camp-on-mount-athos/ (accessed on 16 November 2020). Yet, the landscape is not as "authentic" as presented to visitor-pilgrims, as it has transformed several times in the past, mainly because of the great fires of 1580, 1622, 1891 (Eleseos and Papaghiannis 1994, p. 48), and most recently, in August 1990, and February 2004, and is constantly changing. Eleseos and Papaghiannis identified as the main ecological problem the desertification of the land, especially at the southern areas of the peninsula, because of the great fire of 1990 and over-exportation of wood (Ibid, pp. 51–54). The two monks further observed that "the introduction of telecommunications, water pipes, machines, and electrical generators into the peninsula threat the calmness, form and function of the environment . . . The pollution of the space from concrete and liquid waste could be out of control" (Ibid, 43, *my translation* from Greek).

highly organized environment. The presence of guests was limited to the guesthouse in the eastern wing of the monastery, the Chapel of the Holy Girdle of Mary (the item that most visitors travel to venerate), the Refectory and the *Catholicon* (main church) in the center of the monastic complex, and the northern wing of the guesthouse (dormitory). The rooms are decorated with warm red carpets, and beautiful and expensive icons of saints and/or Greek national heroes. The first and second floors have up to eight beds. This area is reserved for common visitors. The third floor is far more luxurious, with big rooms of single or two beds reserved for the rich and powerful "friends" of the monastery, that is, mainly donators, politicians, and Archimandrites. At the top, where my room was situated, was used for visitors who planned to stay for longer periods and possible candidates. All rooms have electricity and hot water throughout the day produced by two generators. The deacons on duty are responsible for changing the bed sheets and pillows on a daily basis. By contrast, the cells of the monks are of a much poorer condition—without electricity or running water situated at the southern wing of the monastery clearly separated from the visitors' area. Visitors are not allowed to access that area, so that they do not disturb the private and traditional way of life of the monks.

The presence of visitors in the daily life of the monasteries is a major challenge to the monks and their traditional obligation, associated with the landscape's sacredness, to offer hospitality. Many visitors, especially newcomers, do not behave like "pilgrims" because of their tourist mentality and ignorance of the rules associated with the *Avaton* and/or ritualized daily life. For this reason, the elders have established a set of rules on how to accommodate visitors (affirmed in the Charter of Athos of 1926) and reinvented traditional monastic roles, such as those of the guest-master (*"archontaris"*) and sacristan (*"vemataris"*) according to the visitors' present needs. Both are responsible in welcoming the visitors, setting the timetable and rules of conduct and contact while staying in the monastery, and then introducing them to the monastery's history and tradition, including an exhibition of the holy relics.

The rules demonstrate an economic quality with its own "logic" of "economic" restrain, from food abstinence in the rooms, to the prohibitions of wearing a t-shirt, talking on mobile phones, taking pictures of the monastery without the permission of the Abbot or swimming—as this is not meant to be a touristic place like the rest of Chalkidiki. Under the omnipotent gaze of Mary, visitors rarely break the rules, and if so, they get blacklisted by the Athonian authorities and are not allowed to revisit the place for a certain period. In this context, such restrictions exhibit another sense of "economy" (from the Greek word *"ecos"* and *"nomos"* meaning "Law of the House"): economy in how one is dressed, and economy in how one uses technology inside the monasteries. In this sense, visitors are integral in the economical organization of the monasteries. Visitors buy copies of holy items, books and CDs from the monastery's gift shop usually situated next to the main gate or inside the guesthouse. In exchange for their staying, some visitors who spend longer periods there engage in voluntary work, for example, packaging handmade products, rosaries and candles, CDs with the choir, and books on the lives of charismatic elders, in order to distribute them to their local churches and religious shops outside Athos. Such economic arrangements mediate between the secular world and prohibitions associated with the land, guaranteeing the naturalized moral order of the monasteries within the wider landscape reinforced by a collective feeling that Mary is constantly watching them.

In the same way they divide the spaces of the monastic buildings in terms of accessibility, the monks keep the monastic and secular worlds separated via two parallel internal hierarchical structures: an informal spiritual hierarchy of elders and deacons running parallel to more institutionalized forms of rank of priests and priest-deacons (Sarris 2000, pp. 8–9; Paganopoulos 2009, pp. 364–66). The structural separation of the public (daily activities including accommodation of visitors and addressing the material needs of the monastery) from the monks' private life (spiritual activities that take place in the night addressing the spiritual needs of the monastic self) remain integral in the monasteries' focus in accepting and accommodating their visitors. These economic compromises are integral in the traditional life of the community. The accommodation of visitors within the strict hierarchy of the monastery is a precondition in order to include them as "pilgrims"—regardless of their background or degree of faith—by keeping their presence separate from the monks and their working and private areas, in terms of rules and prohibitions, accessibility, and manner of conduct. This form of control is a precondition for entering a monastery in the first place, including the limited number of visitors (it takes up to three months to gain a permission to enter the republic by the Athonian authorities)[11]: "The number of visitors permitted on the Holy Mountain at any time is tightly restricted and all visitors are, by definition, pilgrims. Whatever your reason for visiting them, the monks will welcome you as a pilgrim".[12] As Gothóni exclaims, "while many visitors on their inward journey admit they are just taking a look, when they return, they realize they have made a pilgrimage" (Gothóni 1994, p. 179, cited by Andriotis 2009, p. 79). So, how does this metamorphosis of "tourists" into "pilgrims" take place (in order for them to be accepted by the community)?

6. Personal Services

In my discussions with visitors, many of them visited Athos not only for sightseeing or tasting Byzantine life, but also for personal reasons, in the lookout for a "spiritual father" ("*pneumatikos pateras*") who would help them in matters over their personal life outside Athos (Paganopoulos 2009, pp. 375–77, and also in Andriotis' example of "influential authenticity" 2011, pp. 1626–7). In the words of Abbot Ephraim of Vatopaidi, one "important" service the monks offer today is giving visitors advice regarding issues in their personal secular life outside Athos: "When lay people visit the monastery, they take away the example of the monks or nuns, their words, their silence, their prayer, their whole attitude and religious disposition. These lay people take leaven, they are nourished spiritually and recharge their batteries [. . .] Then when they return to their family 'in a different form,' the other members will also benefit spiritually". [13] In this context, the monks act as "brokers of the tourist experience" (as in Zatori et al. 2018, p. 112) between the secular and monastic worlds. Their mediatory role allows them to both contain the "authentic experience" but also influence their visitors beyond their visit ("influential authenticity", as in Andriotis 2011, p. 1626). This close engagement with the visitors takes place via personalized services and cathartic practices of faith, such as confession. The rite has both a cathartic and a supervising role, as, on the one hand, through the rite each visitor is cleansed in order to be allowed to stay as a means of entering the hierarchy and communal rule, and on the other, by allowing the impersonal institution of the "Monastery" to enter into the personal life of each visitor. This exchange of private worlds (monastic and personal) is necessary, especially for those who wish to stay for longer periods in a monastery, as a means of allowing and accepting them in the monastery's daily life. Furthermore, the request to confess on arrival is necessary for allowing them to become a *pilgrim*, thus drawing a line between those who have confessed and can stay longer and those *tourists* who have to leave after a few days defining the quality of service they receive.

An important aspect of confession is that the spiritual fathers have to show "economy" towards the visitors by considering each one's background. The "spiritual fathers" ("*pneumatikoi pateres*") cannot ask of a first-time visitor the same attention or show the same strictness as they do to someone who is a more frequent visitor. In this "economical" manner, the spiritual fathers reinvented confessions according to our times and the needs of their visitors as a kind of psychoanalytical session, making them integral and relevant to

11 "A maximum of 120 Orthodox Christian visitors are allowed per day, whereas foreigners of other religious affiliations are limited to 14 per day. These limits do not include persons that have explicit invitations from the monasteries." From "Visiting Mount Athos" in *Mount Athos: The Holy Mountain* official website Available online: http://www.macedonian-heritage.gr/Athos/General/Visiting.html (accessed on 16 March 2021).

12 In Greek, the word "pilgrims", i.e., "*proskynites*" literary means "worshippers. Available online: https://athosfriends.org/pilgrims-guide/ (accessed on 11 November 2020).

13 "Elder Ephraim of Mt. Athos speaks with TNH" (10 February 2018). Available online: https://www.thenationalherald.com/archive_church/arthro/elder_ephraim_of_mt_athos_speaks_with_tnh-19466/ (accessed on 2 December 2020).

each visitor's secular daily life. During confessions, the priest-monk offers his advice and a set of "exercises" (*askesis*, root to the word "ascetic") focusing on prayer, along with his moral advice on secular issues such as sexual intercourse outside marriage, advice over family issues, etc. Over time, frequent visitors develop a trusting relationship with their "spiritual father" that can be life-changing. Furthermore, the confession made me feel as being supervised by the community as a whole, as my private life became public property through my confession to my "spiritual father", who then confessed his sins to the Abbot. In this hierarchical manner, the monks both allow and supervise the presence of visitors into their hierarchical system. The spiritual father functions as a mediator between the individual's world and the sacred place, one that includes both the monastery and the community as a whole. Vice versa, the process of confession aims to gradually transform the "tourist" to a "pilgrim" within the moralized and naturalized terms determined by the sacredness given to the place. The connection between the self and the community is then affirmed and celebrated with the reception of the Divine Eucharist every Sunday by the whole community in a Durkheimian unificatory manner.

7. Contested Hospitality in Esfigmenou

Five kilometers north of Vatopaidi is the monastery of Esfigmenou, "the last tower of zealots" as it is known on social media. Esfigmenou is a self-proclaimed "zealot" brotherhood that belongs to the "Old Calendarist Church" ("*Palaioimerologites*"), an international sect called the "Authentic Orthodox Christians" whose members follow the "old" Julian calendar as a demonstration of their ultra-Orthodox faith (Moss 2010, pp. 3–59; Sidiropoulos 2000, p. 173; Taft 2013, pp. 23–44; Speake 2002, p. 146, and Demacopoulos 2017, pp. 482–89). The monastery of Esfigmenou is at the center of this network, which includes a number of renegade subgroups, such as the "Russian Orthodox Church Outside Russia" (ROCOR) in the US, as well as more localized far-right religious groups in the Balkans, such as those of "St Basil" and "ELKIS" in Thessaloniki, Greece (Moustakis 1983). Since 1973, a black banner hangs from Esfigmenou's highest tower calling for "Orthodoxy or Death". Since February 2003 (the time I was inside for my fieldwork), the monastery has been put under embargo by the Athonian authority and the police, an embargo that lasts to this day. According to the Eviction Note, issued on 5 February 2003 by the Athonian authorities, the new Zealots of Esfigmenou and their followers "have willingly cut themselves off all Orthodox Churches in order to join the so-called 'Authentic Orthodox Christian' group, and only give Holy Communion to members of the Old Calendarist Church".[14] Other transgressions include not commemorating the Ecumenical Patriarchate in their prayers, and engaging in a persistent political activism with their followers inside and outside Athos.

Inevitably, the circumstances and restrictions imposed by the embargo define both the monastic service the monks offer to their visitors, and vice versa, the quality of visitors the monastery receives according to a specific ideological orientation and the institution's reputation in the Orthodox world. In my discussions with visitors in Esfigmenou, I could not help but notice a kind of "persecution complex" collectively expressed as an "apocalyptic mission" (as in Hall 1997, p. 361). The embargo against the monastery was a personal matter for them, in many ways symbolizing their own life struggle against society and their personal demons inside them. The embargo did not prevent them from climbing over the fence that separates Athos from Greece and illegally walk to the monastery via the thick forest. Ironically, the stricter the embargo becomes, the bigger the reputation of the monastery gets, and consequently, the more visitors it receives. Most of the visitors I met in Esfigmenou were social outcasts stigmatized by poverty, drug addicts and alcoholics trying to rehabilitate from their addiction, illegal immigrants and fugitives from the law, right-wing extremists and followers of the brotherhood's neo-fundamentalist dogma of the

[14] Update issued by the Holy Committee informing pilgrims the reasons for the 2003 embargo (28 January 2003, p. 2, *my translation from Greek*]. *See* also pamphlet *The Truth of the Case of the Occupiers of Esfigmenou.* Mount Athos, Karyes.

"Old Calendarist Church". They had arrived from across the spectrum of the Orthodox world and joined one of the seventeen segregated groups of the monastery according to their nationality and language headed by an elder.

The loose hierarchy of the monastery was also evident in spaces such as the refectory where monks and visitors shared the same tables according to their nationality (rather than sitting separately from the monks as in the other monasteries). No confessions or frequent reception of the Holy Communion were offered to visitors, since the "Old Calendarist" dogma considers frequent repetition of practices and values such as of obedience as a means of keeping visitors under the control of the "Papic" West (sic). In open discussions with visitors, the monks of Esfigmenou often used the neighboring monastery of Vatopaidi as an example of "betraying" monks who have distorted the "authenticity" of Athonian life threatening to distort the "sacred tradition" from the inside: *"They will even allow women to visit with their families and make this place a hotel like the rest of Chalkidiki"* one Elder of Esfigmenou exclaimed to me.[15] Instead, the type of exchanges in Esfigmenou exclude money. During the embargo, visitors supported the monks' ideological cause in numbers, bringing with them food, medicine and supplies. The active involvement of the visitors of Esfigmenou in the monastery's daily life was essential for the survival of the monks. Due to the embargo, visitors in Esfigmenou can stay indefinitely, as long as they take the oath to the "Old Calendarist" dogma (instead of confession).

Hence, while in Vatopaidi visitors transformed into "pilgrims" via reinvented practices of faith, such as confession, in Esfigmenou visitors actively participate in daily life by following and fighting for the monastery's ultra-Orthodox ideology inside and outside Athos, and against the Athonian authorities and the Ecumenical Patriarchate. However, for the Vatopaidians, the over-emphasis of Esfigmenou on poverty and struggle undermined the "spiritual life" of Athos and its guarantor, the authority of the Ecumenical Patriarchate. They see Esfigmenou's passionate engagement with the dogma of "True Faith" as against the teachings of the Fathers that promote "silence" ("Hesychasm") and peaceful harmony within the self, as well as towards others and the natural environment. The Vatopaidians also believe that it is their main responsibility to preserve and develop the monastic institutions—for which they need funding and influence outside Athos. By refusing to engage with the rest of the Athonian community, the monks of Esfigmenou *"let their monastery (Esfigmenou) rot"* one Elder of Vatopaidi told me. He continued: *"You know what they (monks of Esfigmenou) do? They sent their visitors dressed up in poor clothes to ask for water, because they already know that we will not offer them because they are all a sect. Then they go around on Athos and in Thessaloniki and complain that we do not honour hospitality."* (Paganopoulos 2012, pp. 273–74, Fieldwork Notes). Here then, the traditional monastic value of *"filoponia"*, meaning "friend-with-pain", which both monks and their followers publicly exhibit in Esfigmenou, is practiced the wrong way because of its excessive use as a way to publicly demonstrate "True Faith".

It is important to note that from the Vatopaidian perspective, it is not the value or the practice that are wrong, but the way they are interpreted and the *style* of performance (Vatopaidian spiritual setting for visitors to renew themselves versus Esfigmenou's setting offering to visiting pilgrims the means for self-martyrdom against the world). This difference in style reflects upon the differences in customization and the way the monks relate to their visitors, and vice versa, the way the visitors relate to the monks. While in Vatopaidi, the processes of interactive customization and male bonding between monks and visitors

[15] The prohibition of women remains a topic of contestation and discussions in the European Parliament over the abolition of the *Avaton* on the basis of the Charter of Fundamental Rights of the European Union, as per article 23 regarding the "Equality Between Men and Women". In 2003, a number of members of the European Parliament raised the issue, including Maria Izquierdo Rojo's (PSE) written question P-0556/03 to the Commission (20/2/2003) published the Official Journal of the EU 2004/C 58 E/023, and Anna Karamanou's and the Chairperson of the Committee of Women's Rights and Equal Opportunities' report in July 2003 [See Swiebel and Rojo reports 2003]. See also Corrigendum to Directive 2004/58/EC of the European Parliament and of the Council of 29 April 2004 on the right of citizens of the Union and their family members to move and reside freely within the territory of the Member States amending Regulation (EEC) No 1612/68 and repealing Directives 64/221/EEC, 68/360/EEC, 72/194/EEC, 73/148/EEC, 75/34/EEC, 75/35/EEC, 90/364/EEC 90/365/EEC and 93/96/EEC (Official Journal of the European Union L 158 of 30 April 2004), and L 229/38 EN Official Journal of the European Union 29.6.2004 'RIGHT OF EXIT AND ENTRY to all Union citizens' (CHAPTER II: Article 4).

take place in "spiritual" passionless ways (confession, private prayer, contemplation), in Esfigmenou, customization is marked by poverty and the public and passionate struggle against both the body and the world. Accordingly, catharsis in Vatopaidi is a private and "spiritual" matter of the soul, but in Esfigmenou is a public matter of struggle against the flesh in a zealot way. The way the monks of each institution relate to their visitors (spiritual versus zealot manner) defines their customer service in terms of the quality of "authentic experience" they are offering, as part of the *specific* vocation of their monastery in the Orthodox world.

8. Conclusions

The material of this paper empirically compared the customization and involvement experience of visitors in two rival monasteries of Athos, demonstrating their opposite understandings regarding the role and means of offering monastic hospitality to the world. The different quality of services offered to, and opposite styles of interacting with visitors, revealed the heterogeneity in the interpretation and practice of hospitality in each monastery, in direct relation with their external vocation in the Orthodox world. In these terms, the paper compared the contrasting internal regimes of Vatopaidi and Esfigmenou as spiritual and ideological environments, respectively. Accordingly, the paper discussed how the monks of each monastery relate to their visitors in their own idiosyncratic manner. he comparison qualitatively unpacked the third element that constitutes the concept of "staged authenticity" (objective, constructive, and existential) in its spiritual and ideological components. It investigated how the claim to "authenticity" is an open arena of contestation and negotiation regarding appropriations of the past in the present time, such as the emergence of the phenomenon of pilgrimage tourism. By contextualizing the pilgrimage tourism of Athos within the world system, the paper further showed how traditional customization has rapidly evolved into a variety of styles and services offered by the monks to visitors, in relation to the global phenomena of re-enchantment with sacred landscapes and emergence of the personalized market of faith.

Funding: This research received no external funding.

Institutional Review Board Statement: Ethical review and approval were waived for this study, due to granted anonymity, transparency in fieldwork conduct, and active involvement of informants who kindly offered material and participated in this research.

Informed Consent Statement: Informed consent was obtained from all subjects involved in the study.

Data Availability Statement: The data presented in this study are openly available in Goldsmiths Research online, under License Creative Commons Attribution Non-commercial No Derivatives, at http://research.gold.ac.uk/id/eprint/6537/ (accessed on 23 March 2021). Item ID: 6537.

Conflicts of Interest: The author declares no conflict of interest.

References

Andriotis, Konstantinos. 2009. Sacred Site Experience—A Phenomenological Study. *Annals of Tourism Research* 36: 64–84. [CrossRef]

Andriotis, Konstantinos. 2011. Genres of Heritage Authenticity. Denotations from a Pilgrimage Landscape. *Annals of Tourism Research* 38: 1613–33. [CrossRef]

Badone, Elen, and Sharon Roseman, eds. 2004. *Intersecting Journeys: The Anthropology of Pilgrimage and Tourism*. Chicago: Illinois UP.

Bauman, Zygmunt. 1998. Postmodern Religion? In *Religion, Modernity and Postmodernity*. Edited by Paul Heelas, David Martin and Paul Morris. Massachusetts: Blackwell Publishers, pp. 55–78.

Bax, Max. 1983. US Catholics and THEM Catholics in Dutch Brabant: The Dialectics of a Religious Factional Process. *Anthropological Quarterly* 4: 167–78. [CrossRef]

Bax, Max. 1990. The Madonna of Medjugorgje. *Anthropological Quarterly* 2: 63–75. [CrossRef]

Belhassen, Yavin, Kellee Caton, and William Stewart. 2008. The search for authenticity in the pilgrim experience. *Annals of Tourism Research* 35: 668–89. [CrossRef]

Bruner, Edward. 2005. *Culture on Tour: Ethnographies of Travel*. Chicago: University of Chicago.

Cohen, Erik. 1988. Authenticity and Commoditization in Tourism. *Annals of Tourism Research* 15: 371–86. [CrossRef]

Comaroff, Jean, and Jean Comaroff. 2000. Millennial Capitalism: First thoughts on a Second Coming. *Public Culture* 12: 291–343. [CrossRef]
Demacopoulos, George. 2017. 'Traditional Orthodoxy' as a Postcolonial Movement. *The Journal of Religion* 97: 475–99. [CrossRef]
Durkheim, Emile. 1995. *The Elementary Forms of Religious Life*. Translated by Karen E. Fields. New York: Free Press. First published 1912.
Eade, John, and Michael Sallnow. 1991. *Contesting the Sacred: The Anthropology of Christian Pilgrimage*. Chicago: University of Illinois.
Eleseos, Priest-Monk, and Priest-Monk Papaghiannis. 1994. *Natural environment and monasticism: Preserving Tradition in the Holy Mount*. Athens: Goulandri.
Gilmore, James, and Pine Joseph. 2007. *Authenticity: What Consumers Really Want?* Boston: Harvard Business School Press.
Gothóni, René. 1994. *Paradise within Reach: Monasticism and Pilgrimage on Mt Athos*. Helsinki: Helsinki University Press.
Hall, John. 1997. Apocalypse at Jonestown. In *Magic, Witchcraft, and Religion*, 4th ed. Edited by Arthur Lehmann and James Myers. London and Toronto: Mayfield, pp. 353–64.
Heelas, Paul, Scott Lash, and Paul Morris, eds. 1996. *Detraditionalization*. Massachusetts and Oxford: Blackwell.
Heelas, Paul, Woodhead Linda, Seel Benjamin, Szerszynski Bronislaw, and Tustin Karin. 2005. *Why Religion Is Giving Way to Spirituality*. Maiden, Oxford and Victoria: Blackwell.
Herzfeld, Michael. 1985. *The Poetics of Manhood*. Princeton: Princeton UP.
Herzfeld, Michael. 1997. *Cultural Intimacy: Social Poetics in the Nation-State*. New York: Routledge.
Iossifides, Marina. 1991. Sisters in christ: Metaphors of kinship among greek nuns. In *Contested Identities: Gender and Kinship in Modern Greece*. Edited by Peter Loizos and Evthymios Papataxiarchis. Princeton: Princeton UP, pp. 135–55.
Kim, Hyounggon, and Tazim Jamal. 2007. Touristic quest for existential authenticity. *Annals of Tourism Research* 34: 181–201. [CrossRef]
Kim, Bona, Seongseop Kim, and Brian King. 2020. Religious Tourism Studies: Evolution, progress, and future prospects. *Tourism Recreation Research* 45: 185–203. [CrossRef]
Kotsi, Filareti. 1999. The enchantment of a floating pilgrimage. The case of Mount Athos, Greece. *Vrijetijdstudies* 17: 5–20.
Kotsi, Filareti. 2012. Mount Athos: Development policies for short term religious tourism. *International Journal of Tourism Anthropology* 2: 149–63. [CrossRef]
Loizos, Peter, and Evthmios Papataxiarchis, eds 1991. Introduction. In *Contested Identities: Gender and Kinship in Modern Greece*. Princeton: Princeton UP, pp. 3–26.
MacCannell, Dean. 1973. Staged authenticity: Arrangements of social space in tourist settings. *American Journal of Sociology* 79: 589–603. [CrossRef]
Malavakis, Nikos. 1999. *Βυζαντινολόγιο-Λεξικό Εκκλησιαστικών και Θρησκευτικών όρων*. Athens: Astir.
Morinis, Alan, ed. 1992. *Sacred Journeys: The Anthropology of Pilgrimage*. Westport and London: Greenwood Press.
Moss, Vladimir. 2010. Thirty Years of Trial: The True Orthodox Christians of Greece, 1970–2000. East House, Beech Hill, Mayford, Woking and Surrey: Holy Great Martyr and Healer Panteleimon. Available online: http://www.orthodoxchristianbooks.com/downloads/257_THE_TOC_OF_GREECE_1970_2000.pdf (accessed on 23 March 2021).
Moustakis, Giorgos. 1983. *The Birth of Christian-Fascism in Greece*. Athens: Kaktos.
Olsen, Daniel, and Dallen Timothy. 2006. Tourism and religious journeys. In *Tourism, Religion and Spiritual Journeys*. Edited by Dannel Timothy and Daniel Olsen. Abingdon: Taylor & Francis eLibrary, pp. 1–22.
Paganopoulos, Michelangelo. 2007. Materializations of Faith on Mount Athos. In *Material Worlds*. Edited by Rachel Moffat and Eugene de Klerk. Newcastle: Cambridge Scholars Publishing, pp. 122–3.
Paganopoulos, Michelangelo. 2009. The concept of Athonian Economy in the Monastery of Vatopaidi. *Journal of Cultural Economy* 2: 363–78. [CrossRef]
Paganopoulos, Michelangelo. 2012. The Land of the Virgin: An Ethnography of Monastic Life on Athos. Ph.D. dissertation, Goldsmiths, London. Available online: http://research.gold.ac.uk/6537/ (accessed on 26 October 2019).
Paganopoulos, Michelangelo. 2020. Being and Becoming a Monk on Mount Athos: An Ontological Approach to Relational Monastic Personhood in the "Garden of the Virgin Mary" as a Rite of Passage. *Open Theology* 6: 66–87. [CrossRef]
Papachrysanthou, Dionysia. 1992. *Athonian Monasticism: Principles and Organization*. Athens: National Bank of Greece.
Rejman, Krzysztof, Piotr Maziarz, Cezary Andrzej Kwiatkowski, and Małgorzata Haliniarz. 2016. Religious tourism as a tourism product. *World Scientific News* 57: 562–75. Available online: http://www.worldscientificnews.com/wp-content/uploads/2016/06/WSN-57-2016-562-575.pdf (accessed on 30 November 2020).
Ross, Ellen. 1991. Diversities of Divine Presence: Women's Geography in the Christian Tradition. In *Sacred Places and Profane Spaces: Essays in the Geographies of Judaism, Christianity, and Islam*. Edited by J. Scott and P. Simpson-Housley. New York: Greenwood Press, pp. 93–114.
Sallnow, Michael. 1981. Communitas Reconsidered. *MAN* 16: 163–82. [CrossRef]
Sarris, Marios. 2000. Some Fundamental Organizing Concepts in a Greek Monastic Community on Mt Athos. Ph.D. dissertation, Department of Anthropology, LSE, London, UK.
Selwyn, Tom, ed. 1996. *The Tourist Image: Myths and Myth Making in Tourism*. Chichester: John Wiley & Sons.
Seremetakis, Nadia. 1991. *The Last Word: Women, Death, and Divination in Inner Mani*. Chicago: University of Chicago Press.
Sidiropoulos, Georgios. 2000. *Mount Athos: References in Anthropogeography*. Athens: Kastanioti.

Solomon, Esther, ed. 2021. *Contested Antiquity: Archaeological Heritage and Social Conflict in Modern Greece and Cyprus*. Bloomington: Indiana UP.

Speake, Graham. 2002. *Mount Athos: Renewal in Paradise*. New Haven and London: Yale UP.

Speake, Graham, and Dimitrios Conomos, eds. 2005. *Mount Athos the Sacred Bridge: The Spirituality of the Holy Mountain*. Oxford: Verlag Peter Lang.

Speake, Graham, and René Gothóni, eds. 2008. *The Monastic Magnet: Roads to and from Mount Athos*, 3rd ed. Oxford: Verlag Peter Lang.

Taft, Robert. 2013. Perceptions and Realities in Orthodox-Catholic Relations Today. In *Orthodox Constructions of the West*. Edited by George Demacopoulos and Aristotle Papanikolaou. New York and Oxford: Fordham UP, pp. 23–44.

Tanasyuk, Pavlo, and Chrisanthi Avgerou. 2009. ICT and Religious Tradition: The Case of Mount Athos. Available online: http://</named-content>eprints.lse.ac.uk/35560/ (accessed on 16 November 2020).

Theodossopoulos, Dimitrios. 2013. Laying Claim to Authenticity: Five anthropological dilemmas. *Anthropology Quarterly* 86: 337–60.

Tilson, Donn James. 2005. Religious-Spiritual Tourism and Promotional Campaigning: A Church-State Partnership for St. James and Spain. *Journal of Hospitality & Leisure Marketing* 12: 9–40.

Turner, Victor. 1974. *Dramas, Fields and Metaphors: Symbolic Action in Human Society*. London: Cornell.

Turner, Victor. 1997. *The Ritual Process: Structure and Antistructure*. New Brunswick and London: Aldine Transaction.

Tzanelli, Rodanthi. 2008. *Nation Building and Identity in Europe: The Dialogics of Reciprocity*. Hampshire: Palgrave Macmillan.

Vergotis, Georgios. 1995. Λεξικόν Λειτουργικων και Τελετουργικων Ορων, 3rd ed. Available online: https://anemi.lib.uoc.gr/metadata/2/4/a/metadata-01-0001763.tkl (accessed on 23 March 2021).

Wang, Ning. 1999. Rethinking authenticity in tourism experience. *Annals of Tourism Research* 26: 349–70. [CrossRef]

Wang, Kuo-Yan, Azilah Kasim, and Jing Yu. 2020. Religious Festival Marketing: Distinguishing between devout believers and tourists. *Religions* 11: 413. [CrossRef]

Zatori, Anita, Melanie K. Smith, and Laszlo Puczko. 2018. Experience-involvement, memorability, and authenticity: The service provider's effect on tourist experience. *Tourism Management* 67: 110–26. [CrossRef]

Zhang, Hao, Taeyoung Cho, Huanjiong Wang, and Quansheng Ge. 2018. The Influence of Cross-Cultural Awareness and Tourist Experience on Authenticity, Tourist Satisfaction and Acculturation in World Cultural Heritage Sites of Korea. *Sustainability* 10: 927. Available online: https://ideas.repec.org/a/gam/jsusta/v10y2018i4p927-d137640.html (accessed on 16 March 2021). [CrossRef]

Article

'Biker Revs' on Pilgrimage: Motorbiking Vicars Visiting Sacred Sites

Ruth Dowson

UK Centre for Events Management, School of Events, Tourism and Hospitality Management, Leeds Beckett University, Leeds LS6 3QU, UK; r.dowson@leedsbeckett.ac.uk

Abstract: In April 2014, a new Church of England diocese was instituted, combining three smaller dioceses covering a large area of Yorkshire. To mark the development of this new 'mega-diocese', a group of motorcycling vicars began to meet regularly and undertake 'rides out' across the diocese and further afield. This paper explores research undertaken with these motorbiking priests and their companions. The study followed an ethnographic approach, as the researcher is an ordained clergyperson embedded within the 'Biker Revs' community, though not a biker. The research comprised semi-structured interviews and informal conversations with the Biker Revs over several years. This research investigates the Biker Revs' experiences and motivations for undertaking pilgrimages together, by motorbike. On these performative journeys, the Biker Revs initially visited sacred sites across the United Kingdom. As a basis for comparison, this paper utilizes Michalowski and Dubisch's 2001 iconic ethnographic research on an American motorcycle pilgrimage, to analyze the group's activities. The ordained bikers identified the group as a safe space for clergy, outside their parishes, whilst their spouses recognized the benefits of spending time with 'others like me who understand the pressures of clergy life'. For some participants these pilgrimages provide a religious retreat, as together, they explore sacred landscapes and learn the stories of their holy destinations.

Keywords: pilgrimage; sacred sites; motorbiking; biker; Church of England; retreat; eventization of faith

Citation: Dowson, Ruth. 2021. 'Biker Revs' on Pilgrimage: Motorbiking Vicars Visiting Sacred Sites. *Religions* 12: 148. https://doi.org/10.3390/rel12030148

Academic Editor: Kiran Shinde

Received: 1 February 2021
Accepted: 22 February 2021
Published: 25 February 2021

1. Introduction

This research considers the impacts of a decision by two vicars to respond to an historic change to their local diocese. As background, in April 2014, a unique moment for the Church of England witnessed the merger of three existing Yorkshire dioceses into one, creating a new mega-diocese. Figure 1 below shows a map of this new diocese (Leeds Diocese 2014). The response from these two vicars to the creation of the new entity, was to form a group of motorbike enthusiasts, open to clergy and lay people across the wider geographical area. The group became known as the 'Biker Revs', with the purpose of bringing together lay and ordained people from across the new diocese in a shared activity—motorcycling. Mainly comprising vicars and their spouses, the Biker Revs have become a regular sight in the region, and beyond. Originally named the Diocese of West Yorkshire and the Dales, two years later, it became the Diocese of Leeds (Bogle 2016).

On their first 'ride out', in October 2013, the group visited the cathedrals of the three dioceses that would soon be merged—Ripon, Bradford, and Wakefield. In addition to an average of four local day rides out each year, the group has also undertaken five longer motorbike pilgrimages (each five to six days in length). For their first pilgrimage, in April 2015, the Biker Revs visited the island of Iona, off the coast of western Scotland, widely regarded as the cradle of Christianity in the British Isles. In June 2016, for their second journey, the group travelled to Ballycastle in Northern Ireland, home of the peace and reconciliation Corrymeela Community (n.d.). In June 2017, continuing the theme of visiting places where the national saints of the United Kingdom are remembered, the Biker Revs rode to St David's in Wales.

Figure 1. Map of the new Diocese of West Yorkshire and the Dales (2014) (Leeds Diocese 2014).

In October 2017 a select group ventured further afield, undertaking the Santiago de Compostela pilgrimage route by motorbike. In June 2019, they spent a week biking around the North Yorkshire coast, visiting Holy Island and the Lindisfarne Priory. In 2020 the group's plans were curtailed by COVID-19 restrictions. The map of the United Kingdom in Figure 2 below shows the longer trips undertaken by the Biker Revs. In the Church of England, all clergy, whether paid or unpaid, full-time or part-time, are expected to go on 'retreat' for a week every year. In addition, after seven to ten years following ordination, clergy may apply for a three-month sabbatical, and many include an extended time for reflection on pilgrimage to a sacred destination. The originality of this research lies in its analysis of pilgrimages undertaken by a group of clergy by motorbike.

The specific goal for the formation of the group was to build relationships between people from (clergy and non-clergy) motorbike enthusiasts from across the new diocese. The approach the group took to achieve this aim was to arrange day rides out and longer five- or six-day pilgrimages to sacred places, initially in the UK. This research investigates the motivations for these pilgrimages (a conventional form of religious tourism) of the Yorkshire Biker Revs as they visit sacred sites, and analyzes the group's activities and experiences. It takes as a theoretical focus, the iconic ethnographic research undertaken by Michalowski and Dubisch (2001) on the motorbike pilgrimage route that reconnects American ex-servicemen with their Vietnam War historical roots, crossing the United States from California to Washington DC. It incorporates reflections on motorcycle pilgrimage by Dubisch (2004, 2008), and Dubisch and Winkelman (2005).

Figure 2. Map of the United Kingdom showing Biker Revs trips (adapted from Google Maps).

This paper also aims to contribute to an understanding of the concept of the eventization of faith, which has emerged from research by Pfadenhauer (2010) on the Roman Catholic World Youth Day. Pfadenhauer's analysis identified connections between the developing popularity of secular music festivals, that had been adapted by the church as a form of marketing to reach young people. Further research has identified that this is a more complex concept that transcends marketing and impacts on a wide range of church-related activities. The Biker Revs have joined in the proliferation of events-based leisure activities through their organization of regular day rides out and longer pilgrimage trips, as they build relationships by visiting sacred places together. In previous research, Dowson et al. (2019) suggest that pilgrimage is a contributing activity to the eventization of faith, and consider the appropriateness of the use of church buildings for non-sacred events and activities (such as riding motorcycles into a cathedral, as shown in Figure 3, below).

Figure 3. Biker Revs on their first 'ride out', inside Bradford Cathedral, October 2013. (Photograph Credit: Rev Stephen Kelly).

2. Literature Review

This section considers key areas of academic literature as it relates to my research questions, on the purpose and impacts of the group's engagement with pilgrimage and motorcycling, both as a leisure activity and as a chosen mode of travel. The literature is applied to the context of the research subjects, and to the activities of these Biker Revs as they perform their pilgrimage journeys. Figure 3 below shows three of the Biker Revs on their motorcycles in the sacred space of Bradford Cathedral on their first ride out in October 2013.

2.1. Pilgrimage

The ancient concept of pilgrimage is found across major world religions; it focuses on a journey made towards reaching a sacred goal (Leppäkari and Griffin 2017). Historically, the journey is physical, often made through dangerous territory and strange lands, but has spiritual and other internal rewards. Pilgrimages vary even in their historical religious origins and purposes. Norman (2011) asserts the importance of the physical journey in addition to the arrival at a sacred destination. The characteristics and purposes of the pilgrimage journey are relevant (Cassar and Munro 2016), along with the eventual pilgrimage destination (Scriven 2014), and the emergence of the feeling of belonging and shared identity (Merkel 2015), or 'communitas' (Turner and Turner 1978), through shared activities. However, visiting pilgrimage sites such as war memorials (Shackley 2001), or places that contain sacred objects (du Cros and McKercher 2015), extends the diversity of what is conceived of as sacred. In the same way, the mode of travel has diversified, and this is reflected in Michalowski and Dubisch's (2001) work on the motorbiking pilgrimage across the USA.

2.1.1. Pilgrim–Tourist

Theoretical explorations of the pilgrimage experience and the pilgrim journey inform a critical view of the pilgrim-tourist dichotomy, explored in depth by Cassar and Munro (2016). To what extent is a pilgrim also a tourist, and to what extent does the tourist become a pilgrim (Fladmark 1998; Raj and Griffin 2015) have been long-discussed. Sturm (2015, p. 237) addresses this type of relationship in a different context (cricket), concluding that there is a 'fluidity' between the roles of cricket spectator and tourist. This idea is applied to the pilgrimages of motorbiking vicars. Similarly, Dowson et al. (2019) analyze the role changes experienced by academics observing a tourist performance of the whirling dervishes in Konya, Turkey during a religious tourism conference. In this instance, the participants found themselves at different points on a pilgrim–tourist continuum. However, many observed the event as academics, and some also as worshipers. This positioning was found to vary at different points in the event (Dowson et al. 2019, pp. 3–4).

2.1.2. The Journey and the Destination—Motivations

Many pilgrims begin their journeys as individuals, setting out for a distant destination. However, as they journey, they meet others and bond together. Their motivations may be with different purposes in mind, including devotion, a search for miracles, ritual, and quest (Norman 2011; Kujawa 2017). Fladmark's definition of spiritual motivations for pilgrimage is relevant: "the innate desire of the human heart to visit places made holy by the birth, life or death of a god or prophet" (Fladmark 1998, p. 45). The Biker Revs' longer rides are such pilgrimages, connecting with British patron saints. Kaell argues that there is a specific Western perspective to pilgrimage, aiming to achieve "something 'good' in the lives of those who undertake it" (Kaell 2016, p. 6). The search for authentic, life-changing experiences through the pilgrimage journey (Abad-Galzacorta and Guereño-Omil 2016) also applies to the Biker Revs. There is growing popularity of pilgrimage by travelling a specific route, from the Camino de Santiago that originated in the late Middle Ages, to the Ignatian Way, a newly founded 21st century pilgrimage route (Abad-Galzacorta and Guereño-Omil 2016). As the Run for the Wall (Dubisch 2008) provides a transformational space for Vietnam war veterans, so the Biker Revs' pilgrimages engage participants in opportunities for spiritual and physical renewal.

Michalowski and Dubisch summarise pilgrimage as combining "a ritual journey with seriousness of purpose, ending in the arrival at a sacred goal" (Michalowski and Dubisch 2001, p. 15). However, there is a symbolic role within the motorcycle journey, expressed as uniting biker pilgrims: "riding together ... joins the participants ... in a common brotherhood of the road" (ibid). At the same time, bikers are set apart from those other travelers who use "more ordinary (and more comfortable) means" of transport (ibid). The Biker Revs experience forms of this strong connection, as they lead or follow within the group. On occasional road traffic accidents involving Biker Revs on trips and rides out, the care shown is evident in the efforts made to ensure everyone is watched over during the journey. Further evidence for this caring element in the Biker Revs' motivations is found in a social media post by one of the founders. His rationale for the group's formation came from Richard Baxter, a 17th-century Puritan minister in the Church of England:

One of my pastoral and priestly heroes: 'Ministers have need of one another and must improve the gifts of God in one another; and the self-sufficient are the most deficient, and commonly proud and empty.' A good enough reason for a biker group!

2.1.3. Liminality

The concept of liminality is relevant to this study in a number of ways: liminality of physical space, liminality of space-time, the liminal experience of traveler, tourist and pilgrim, and the liminality of being both clergy and biker. Turner and Turner (1978) researched the liminal experience of travelers concluding that they experience a liminal space simply by travelling away from their daily lives. Norman (2011) highlights this contrast with everyday life, considering that tourism also provides such an experience. For religious or spiritual tourism, Norman (2011) suggests that this liminal element is explicitly different from the everyday, whether differences are spiritual, religious, or not. However, Norman (2011) concludes that pilgrimage offers availability and access to different kinds of spiritual and religious experiences compared to those at home. Dubisch observes that a pilgrimage undertaken by motorbike "distinguishes their journey from other mundane trips, and particularly from purely recreational motorcycle rides" (Dubisch 2004, p. 114). Whilst most Biker Revs are 'recreational' hobbyists, some rely on their bikes for day-to-day transport.

Motorbiking provides a space-time for bikers; as they travel, they are isolated from other riders, yet still connected to each other (Michalowski and Dubisch 2001). Even riding as a biker with a pillion passenger, the two are separated throughout the journey. Amongst the Biker Revs, several couples have an in-helmet intercom system to enable them to talk whilst riding. For one clergy couple, this was a short-term intervention. Both introverts, they valued the time and space to reflect during riding time, as they left "ordinary life" (Michalowski and Dubisch 2001, p. 166) behind. This is frequently viewed as retreat time, in silent contemplation of the landscapes they pass through (Norman 2011). Travel time removes the traveler from the usual distractions of life; traveling by motorbike more so (Dubisch 2004; Norman 2011; Scriven 2014). Motorbiking offers the opportunity for reflection and self-transformation (Michalowski and Dubisch 2001, p. 166), a vital time to see possibilities for change. Traveling by motorbike is not the most comfortable experience, especially in poor weather. Sitting in the same position for hours on end causes physical discomfort, requires perseverance, and regular stops. Little surprise then, that bikers inhabit a space on the edge, choosing physical hardship as well as thrills for their mode of transport, as they visit sacred sites, "spatially liminal locations separated from the normal world by virtue of their sacrality" (Norman 2011, p. 95). Whilst on rides out or on pilgrimage, the Biker Revs become a "transient, yet liminally-bonded community" (Norman 2011, p. 31), as they seek to reconnect with the other and with each other (Michalowski and Dubisch 2001). This liminal state is characterized by what Turner and Turner name "communitas" (Turner and Turner 1978), as the bikers develop a sense of belonging to one another during their time together.

2.1.4. Communitas

Turner and Turner (1978) introduced the new concept of communitas through their anthropological studies. As people engage in liminal activities together, a social relationship emerges, outside of normal hierarchies and bonds (Moore 2001). The social aspects of a ride out can lead to communitas—but a longer motorcycling pilgrimage is more likely to nurture this outcome. Esposito (2010) considers etymological differences between the Latin 'communitas'—an expression of cooperative obligation in common with others–and the Greek 'koinonia' within its theological context—of joining together through participation in the Eucharist. The Biker Revs develop communitas on their sacred journeys, and they experience koinonia as they share in the sacred ritual of Holy Communion, whether on a beach on Iona or in Northern Ireland.

Michalowski and Dubisch note the impact of communitas on veterans as they bike their way from West to East across the United States, as "the ordinary divisions and distinctions of society are dissolved, at least for a time, and participants enter a state of equality and share experience" (Michalowski and Dubisch 2001, p. 166). For Biker Revs, distinctions between clergy and lay bikers fall away when riding out together; archdeacons ride alongside newly ordained curates, vicars accompany lay people.

2.2. Places of Pilgrimage

The Biker Revs decided to visit sacred places representing the patron saints of the four nations within the British Isles. These trips express the unity (Power 2006) of the nations, symbolizing for the group's founders, the unity of the new mega-diocese of Leeds. Such spiritual destinations offer a sense of place, providing a "spiritual homeland" (Power 2006, p. 35) for pilgrims, and a sense of belonging. The Biker Revs also sought this sense of belonging, searching for a feeling of home in the new diocese and of belonging to each other, and a spiritual connection to their choices of sacred destinations. These pilgrimage destinations are all places of worship (Power 2006, p. 41), an important factor for clergy, enabling them to worship together as equal members of a congregation, in contrast to their everyday calling as leaders in the church. The destinations chosen by the Biker Revs are predominantly centers for Celtic spirituality, "thin places" (Béres 2012; Power 2006) connecting heaven and earth, the spiritual realm and the earthly realm, providing opportunities to connect with the divine.

Iona is known as the "cradle of Christianity" in Scotland but was equally the well-spring of Christian worship in the wider British Isles. Visitors and pilgrims to the island can stay in this monastic community, with shared meals and time to relax together, as experienced by the Biker Revs.

The Holy Island of Lindisfarne in Northumbria is recognized as the "cradle of English Christianity" (Symes 2009, p. 10), and resulted in the production of the most beautiful artwork treasures such as the Lindisfarne Gospels, and the promulgation of worship in the vernacular (English), rather than Latin (Symes 2009).

Power notes that "Corrymeela does not have the same ancient sense of place that make Iona and Holy Island ... significant" (Power 2006, p. 47). However, it is a place of Christian forgiveness and reconciliation, established in the 1960s to promote theological and political values to emulate the Iona community, in response to the political and social context of Northern Ireland (Collins 2000; Power 2006).

For their trip to Wales, the Biker Revs stayed in the countryside, visiting St David's Cathedral on a day ride out (Power 2006). Accommodation and catering arrangements for this trip differed from previous tours, as the group provided their own catering. The photograph in Figure 4 below shows members of the group relaxing outside in after-dinner sunshine.

Meal by meal, the group bonded on their Welsh trip in new ways, forming deeper relationships in the kitchen as they prepared food together. As with all pilgrimages, the Biker Revs return "to the quotidien life" (Scriven 2014, p. 252), renewed after completing acts of devotion in sacred places.

Figure 4. Biker Revs on their Welsh pilgrimage, June 2017. (Photograph Credit: Matthew Dowson).

2.3. Motorcycle Research

The International Journal of Motorcycle Studies considers cultural aspects of motorcycling, from experiential perspectives, technical aspects and the use of motorcycles in film and literature. Relevant publications outline the range of motorcycle writing and research (Alford and Ferriss 2006) providing autoethnographic studies. e.g., the social identity of a military motorcycle club (Wiggen 2019). Literature on Christian motorbikers has several perspectives: retelling personal experiences (Stillman 2018), and exploring stories of Christian motorcyclers (Remsberg 2000; The 59 Club n.d.). However, of more interest is the work of American researchers, Michalowski and Dubisch (2001), on the pilgrimage aspects of a longer ride supplemented by Dubisch's other works (Dubisch 2004, 2008). The underlying assumptions of Michalowski and Dubisch's work support their search for meaning in the aftermath of the Vietnam War, and the roles played by the US military in that conflict, that subsequently impacted on so many veterans' mental and physical wellbeing (Michalowski and Dubisch 2001, p. x).

The concept of motorcycle pilgrims is well-documented by Michalowski and Dubisch (2001), through their participation in an annual pilgrimage by motorbikers, 'Run for the Wall' which has taken place since 1989. This journey begins in California and draws in war veteran riders along the way to the memorial wall commemorating American soldiers who lost their lives in the Vietnam War. A compellingly told narrative, the account explores the motivations and emotional experiences of those seeking healing, making meaning and reconstructing community memory along the route and at the destination, which for many has become a sacred place.

According to Michalowski and Dubisch, the reasons for undertaking pilgrimage include the acquisition of "spiritual benefits", seeking "healing for physical or psychological problems", an opportunity "to honour the holy places of their religious traditions", "to establish or affirm their own religious, cultural or personal identity", and "to express political or social protest" (Michalowski and Dubisch 2001, p. 15). Each day begins with religious rituals, including shared prayers, led by accompanying chaplains. Every stop, during the day or at night, finds local hosts providing free hospitality—accommodation, food, and refreshments. New "deep and sustaining bonds of friendship" (Michalowski and Dubisch 2001, p. 19) are made, and existing attachments renewed.

Michalowski and Dubisch emphasize the differences between this trip and other bike rides out, asserting that they were "not on a pleasure jaunt ... we were pilgrims on a mission" (Michalowski and Dubisch 2001, p. 15). Some of the characteristics of biking pilgrimage match those of pilgrimage through other means, as Michalowski and Dubisch observe that "motorcycle pilgrimages are more than rides ... journeys that change the traveler" (Michalowski and Dubisch 2001, p. 64). Whilst this is ostensibly a secular pilgrimage, it clearly incorporates Christian rituals and other forms of spirituality.

3. Method and Limitations

This paper unashamedly espouses the ethnographic approach, which is common to academic research on churches and their congregations (Scharen 2012; Ward 2012). Ward asserts that "methods of research are never neutral" (Ward 2012, p. 9), and I concur with this view. As a researcher, I am situated within the area that I study. Like Ward (2012), this influences not only the topics that I research, but also what I observe, what I take note of, and what I write. There are benefits to this qualitative ethnographic approach that enable the voice of the research to emerge. In some quarters of the academy, there are still those who question the validity of such research approaches and claim their own objectivity. However, increasingly, in a range of fields, strong considerations of emerging perspectives of authenticity and credibility resist and replace the dominant traditional (quantitative) research discourse of validity and reliability (Frisk and Östersjö 2013).

Of the main methods of qualitative research, this ethnographic study has involved observation (Ward 2012, p. 8) of the group of Biker Revs as it has developed over a period of seven years. The observation method is fundamental to research in the fields of sociology and anthropology, and is the "chosen method to understand another culture" (Silverman 2019, p. 42). Research into church communities also relies on observational studies to "develop a picture of the lived experience" (Ward 2012, p. 8). This study has been based around opportunities to observe a community from inception, as the group of motorbiking vicars has developed strong relational bonds through their pilgrimages and day rides out. As the spouse of one of the initial members, I have been able to connect with the group, even though I do not ride with them. As a result of receiving an annual retreat grant from the diocese, I was able to accompany them on two pilgrimages (to meet them at the destination), to Northern Ireland in 2016, and the following year, to Wales.

The photograph in Figure 5 below shows the Biker Revs group outside the interfaith peace-making Corrymeela Community that hosted their stay in Northern Ireland.

It was on the trip to Northern Ireland in 2016 that I first experienced the sense of belonging to this community, as we shared meals and worshipped together, and explored the area nearby (although my exploration was mostly on foot, as I am not a biker).

This trip also inspired my research, and the following year, having obtained ethical approval for my research, I accompanied the group to Wales. We stayed in self-catering accommodation, where the kitchen became the hub of all interaction, as together, we prepared meals. Silverman (2019) highlights the important role of interviews in qualitative research, in which the responses to open-ended questions can be coded to identify themes. Silverman (2019) and Ward (2012) argue that authenticity is key to qualitative research interviews, rather than the number of interviews that are conducted. The primary research interviews with the Biker Revs explored motivations, views, and experiences in building a sense of kinship and passion for a shared pursuit, that perhaps differentiates bikers from other leisure activities. Figure 6 below summarizes details of the research modes, locations, and participants.

Figure 5. Biker Revs on Pilgrimage, Corrymeela Community, Northern Ireland, June 2016. (Photograph Credit: Stephen Kelly).

Mode of research	Location	Participants
Face to face interview	on Tour: Wales	Individuals / Group / Couples
	not on tour: Bradford	Individuals / Group
Telephone interview	Telephone	Individuals
Questionnaire	Online	Individuals
Participant-Observer	on Tour: N. Ireland, Wales not on tour: Bradford	Researcher

Figure 6. Details of research mode, location and participants.

The interviews were recorded and later transcribed, undertaken in the Welsh countryside, sitting in the lounge or relaxing outside. Some interviews took place in small groups, some with couples together, and some with individuals. Prior to and after the trip, I also undertook interviews by phone to include other members of the group who had not been able to participate in the five-day pilgrimage.

Figure 7 below displays the demographic details of interviewees. Figure 7a shows the split between face to face and telephone interviews, as well as how the interviews were undertaken (individual/couple/group). Figure 7b depicts the numerical gender, seating and role differences of those interviewed (the numbered axis represents the totals of individuals in each grouping).

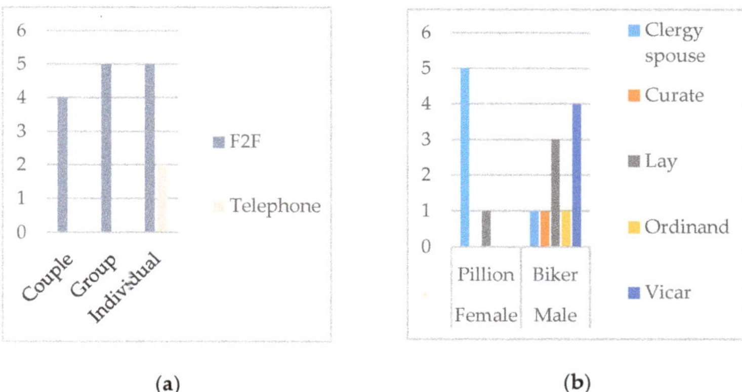

(a) (b)

Figure 7. Demographic research details. (**a**) modes and types of interviews, (**b**) gender, seating and role differences.

On the Wales pilgrimage, it happened that all the bikers were male, and all the pillion riders were female. However, although this is a typical biker pattern, there are now more female bikers in the group (who are all clergy) and on the Northern Ireland trip there was one female biker (who was on honeymoon, having married the weekend prior to the trip). Within the research analysis, there emerge questions of whether there are differences between the views of clergy and non-clergy; stipendiary (full-time paid clergy) and non-stipendiary ordained (unpaid 'volunteer' clergy) participants; the roles of vicars, curates, ordinands, clergy spouses, and lay people; and bikers for whom their bike is their primary mode of transport as opposed to those for whom biking is a hobby—even if the hobbyists have three or more motorbikes in their garage.

My research includes personal reflections on accompanying the Biker Revs (as a non-biker priest, whose non-Rev husband is one of the founder members of the group) on two of their pilgrimage rides, to Wales and Northern Ireland. In this role, I was a pilgrim but not a Biker. Michalowski and Dubisch describe their motorcycle pilgrimage research as ethnography of a cyclic ritual, and their participation as being "motivated by strong feelings that we had to go" (Michalowski and Dubisch 2001, p. 20). This paper's primary research shares that ethnographic perspective, with my role as researcher situated within the group, as a participant-observer reflecting on subjective, personal experiences (Béres 2012; Scharen 2012). In common with the group's founders, I share the characteristics of being ordained, located within the Diocese of Leeds, and, whilst not a biker myself, I am married to a biker who is a member of this group. As a Participant-Observer (Béres 2012; Scharen 2012), and as an ordained person who has accompanied the group on several longer trips, I concur with Michalowski and Dubisch, who observed that

> *We could not separate our personal identities from our professional selves . . . experiences bound us to the other riders and enabled us to understand their cause more than any statement of purpose or ideology from them could ever have done.* (Michalowski and Dubisch 2001, p. 20)

4. Results and Discussion

The research results of interviews with the Biker Revs have been compared with the pilgrimage research outcomes of Michalowski and Dubisch (2001), especially with regard to the identified purpose of the rides, and of group identity. The initial trigger for the Biker Revs group was the creation of a new diocese, bringing together three former dioceses and additional geographic area changes, a unique event in the history of the Church of England. The purpose of setting up this group was to do something that helped both ordained and lay people from across the new diocese to "come together and get to know each other", through a "shared interest", —motorbiking.

The Biker Revs' experience of pilgrimage is applied to the Pilgrim-Tourist dichotomy, as the roles of biker/pillion rider and ordained clergy/spouse of ordained clergy/lay person also impact on the experience of pilgrimage and tourism, as expressed in Figure 8, below. This visually expresses an output of the research, demonstrating differences between the different roles. At the time of the research, all the bikers were male, and most bikers were also clergy. Similarly, all pillion riders were female, and most were clergy spouses. Figure 8 is my attempt to provide a summary interpretation of data from interview transcripts of responses to questions regarding the positioning of participants on the pilgrim–tourist scale (the bold line, left—pilgrim, to right—tourist).

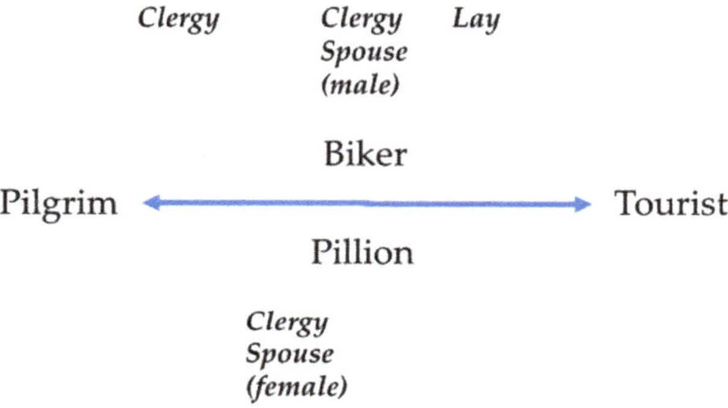

Figure 8. Pilgrim–tourist continuum applied to the Biker Revs.

I have positioned the labels in a way that attempts to represent the grouped positionings on the pilgrim–tourist axis. So, for example, (male) clergy bikers were most likely to define the longer trips as "pilgrimage", and most likely to articulate reasons for motivation and experience that matched the academic definitions of characteristics of pilgrimage, such as being on a journey with a sacred purpose and destination, a sense of shared identity and belonging to the group, and personal spiritual benefits.

Female clergy spouses (all pillion riders) were most articulate on the benefits of being part of the group, also with a spiritual perspective of the experience:

> *It's meeting other people, the companionship and actually the support for each other I think that's built up over the different residentials, particularly. I would say that as a spouse you're able to chat to people–like-minded people without any other distractions, and support each other;*
>
> *Sometimes I can't face going on a retreat because I just don't feel holy enough, I just feel I've had enough of trying to be holy and I don't want any more;*
>
> *I suppose it's being in an atmosphere where you can just be yourself really, there's no expectations as a clergy spouse, particularly you feel as though you have a lot of expectations on you.*

The male clergy spouse and the lay members of the group saw the longer trips as slightly less of a religious pilgrimage and slightly more like a holiday, although all participants acknowledged and valued the religious practices that the group undertook together. This was especially noticeable on the first trip, to Iona, where the group followed the Anglican daily office of Morning Prayer together and celebrated Holy Communion on the beach in Iona. On the trip to Corrymeela, the group joined with the community in their daily acts of worship, a different experience from Iona, but recognized for its spiritual significance. However, in Wales, the group chose to have personal prayer times rather than corporate acts of worship, and this change was appreciated by some, who valued the aspect of being on a religious retreat, which is also a requirement for Anglican clergy. One participant explained that in Wales they would have "felt a duty" to come to Morning Prayer because that is "what's expected", but they appreciated the opportunity to continue their own practice instead.

The American Run for the Wall motorcycle pilgrims and the Biker Revs shared the ritual of fellowship and prayer before setting off each day, and the Biker Revs celebrated Holy Communion on their travels, generally holding daily services (Morning Prayer and/or Compline). They recognized in their motorcycle rides, that kinship was a motivating factor; more than a "common interest", it was a "shared passion". Whether they used their motorbikes as their daily means of travel, or were leisure bikers, each told a story about their spiritual journey as bikers.

One interviewee (now a curate) used his motorbike as the focus for his presentation at the interview for discerning whether he should go forward for ordination training:

Like any other biker, I am passionate about my bike and the possibility of riding the open road on a warm summer's day.

Recognising that historically, bikers were seen as "lawless and aggressive ... biker gangs [are] still known as Hell's Angels or the Devil's Disciples", he was able to develop close (male) friendships through biking together, and could "combine our love of bikes with our love of Christ". Another interviewee recounted the story of two incidents that demonstrated the fearful response that they sometimes experienced as a biker group. The first took place on a day ride out in North Yorkshire as the group arrived and parked up. They asked a passer-by, a woman with two small children, to take a photograph of the group. The woman recoiled, and they all realized that she was petrified, as she grabbed her children, saying, "My husband's at the bottom of the hill, he's coming up". The group leader explained that they were all vicars, so the woman relaxed visibly and took the photos. On another trip the group arrived at a local biker café in North Yorkshire, shortly followed by the arrival of another group of bikers, who approached them. When the Biker Revs said "oh, we're vicars", the new arrivals "backed away and left the café". These incidents indicate the extent to which powerful negative stereotypes of both bikers and vicars can instil fear and trepidation. This connects with the sense of liminality in belonging to both groups—bikers and clergy.

When reflecting on the journey and destination, the motivations of the Biker Revs emerged. As Christians and especially those who are ordained, the idea of visiting a holy place is important, and such destinations are viewed as enabling authentic life-changing experiences. The longer trips, in particular, are viewed as transformational spaces. Several interviewees explained the vital role that the longer trips had played in their lives. For one spouse, facing a difficult work situation, the supportive relationships that were built up in this environment enabled sharing of problems and time for reflection, resulting in a positive decision to change direction. This interviewee had not previously known any of the other Biker Revs members, "it felt a bit daunting at first because I didn't know anybody", but the atmosphere was found to be open and welcoming. Another interviewee reflected on the transformative impact of each of the trips he had participated in. The Iona trip took place just before he was ordained and the Northern Ireland trip was a few weeks before his priesting, providing much needed time to unwind as well as "a lot of thinking time"; trips that were "very significant in where they've fallen in my Christian journey". The Biker Revs' longer trips provide space for physical and spiritual renewal, especially for clergy and their spouses, as well as an opportunity to receive pastoral care outside the parish. Most clergy interviewees and their spouses value the nature of the group as providing a "safe space", where they are able to be themselves. "Friendship" and "fellowship" describe the impact of belonging to the group.

Other aspects of transformation have emerged from being part of the group. One surprising example was the change in attitude towards biking. One clergy spouse had never been on a long bike trip prior to coming to Iona, and so her biker husband had gone on earlier trips with their son. But she had always wanted to visit Iona, so she plucked up her courage and went on the trip. The following year, despite not liking boat travel, she joined the Biker Revs in Northern Ireland, which involved a ferry trip each way. Since that time, their joint holidays have been bike trips, including a trip to Europe by ferry. This was an unlikely transformation, made in part because of the strong relationships formed within the group.

Aspects of liminality emerged throughout the interviews, in terms of the physical space that biking embodies. Biking is a solitary experience, an escape, a time when bikers and pillion riders are able to be alone, even when riding alongside others. Kujawa (2017) notes the focus on internalized reflection during pilgrimage, citing Jung's psychological typology that recognises the greater need for space and "alone-time" of introverts. In the interviews, it became apparent that all but one of the interviewees self-identified as introverts. The time whilst biking, even with a spouse, still incorporates separation from the world and from each other. Despite the availability of two-way intercoms, this communication option is often not employed because of their preference for silence. Clergy undertake training in understanding personality preferences as a matter of course, and many are familiar with the concepts of introversion and extroversion, referring to them in the interviews. In particular, one interviewee values the fact that whilst biking he does not "have to talk to people during the day", noting that:

> *in this group there are a lot of people with the same sort of character as me so it's not a problem if you don't speak to them ... it's not strange.*

Additionally, the shared interest of biking and the common professional role as clergy provide a focus for conversation, as well as practical aspects of the trip, such as meal arrangements and journey logistics.

Longer trips provide a liminal space, as participants are removed from the distractions of life, with time for reflection whilst riding. It is significant that the full-time clergy members of the Biker Revs group found the group as providing a "safe space", where they could "be themselves", outside of the confines of their parishes. For clergy and their spouses, Biker Revs trips provide an escape from the constant pressures of parish life, a time shared with others in similar situations, a place of friendship, sharing, openness and honesty, acceptance, freedom to be who they are, without being called on to help and support others "24/7", and time "without being judged", a place "where vicars can have real friends". This sense of communitas is palpable, as participants relate to each other as equals, outside of normal church hierarchies. Not all Biker Revs are ordained; some have joined the group whilst on the journey to ordination and some are lay people, but on the road, there is no hierarchy. The shared experience of biking, whether bikers use their bikes for day-to-day transport, or are leisure riders, connects group members together with strong supportive bonds. Importantly, the main aim for founding the group, to bring people together from across the three former dioceses, has been successful, embedding a sense of belonging to the new diocese as bikers undertake geographic and spiritual journeys towards their sacred destinations.

Michalowski and Dubisch (2001) provided useful insights for interviewing Biker Revs. The spiritual benefits of the longer trips are evident, along with a sense of identity and belonging to the group, whether bikers or pillion riders. It became apparent that bikers and their pillion passengers had different motivations for participating. For the bikers, their bikes and biking were usually their passion, and whether they use their bikes for practical travel purposes in their daily lives, or as their 'hobby', whilst the perspectives of pillion passengers are notable in their diversity. One pillion-rider clergy spouse admitted that, "I hate speed, but it's a gift to my husband£, an honoring of the depth of their marriage relationship that places the other first. This was echoed by another clergy spouse pillion rider who also expressed biking activity as a "gift" to their spouse. These articulations were especially humbling for me, because as a non-biker spouse of a biker, I recognize that I would not be willing or able to give the same gift to my own (much-loved) spouse. In contrast, another pillion rider was "brought up on a pillion with Dad", so biking became second nature. When asked if the Biker Revs longer trips were "more than a holiday", responses were similar across all interviewees, irrespective of their work roles. Clergy emphasized that this was more of a retreat than a holiday, placing themselves firmly in the pilgrimage category; for them, this is not a pleasure jaunt, but an opportunity to be "in a place with others of a like mind". This phrase "like-minded people" —came up time and time again, especially from clergy, female clergy spouses, and ordinands.

This research has identified some key differences: firstly, that there are two key identifiable groupings: on the one hand, clergy with their spouses, and on the other hand, lay people. Having personally experienced the group dynamics it is possible that for clergy, the Biker Revs might emulate the emotional connections forged in pre-ordination training, which can be particularly strong, as clergy meet up for years afterwards with members of their prayer and study groups from theological college. Longer-term research would ask whether participants develop long-lasting, deep relationships through the Biker Revs activities. However, there is definitely a sense of belonging and a clear acknowledgement of identity in the group as it currently exists. The research identified impacts of personality, gender and role differences (clergy/spouse/lay) on responses. Whilst clergy and spouses value the distinctive nature of the group, another interviewee suggested that it was "an open group" with potential to become a "tool for evangelism". Another contrast in impact was of the type of trip. Day rides out had a different outcome and atmosphere, compared to the longer trips which were viewed more as pilgrimage. There was unexpected learning about the positive well-being impact of the group, especially for vicars and their spouses, and discussions on benefits of the group for male spirituality that is not met by the usual church men's groups and related activities.

5. Conclusions

The research indicates that the interviewee's role in the church impacts on responses; these roles include vicars, curates, ordinands, clergy spouses and lay people. Future research might examine the impact of gender, as at the time of my research, the Biker Revs group only included one female vicar, and all the women rode pillion. Since 2018, there have been more ordained women bikers joining the Biker Revs group. The results of this research suggest that there is a significant relationship between engaging in a shared leisure pursuit (in this case, motorbiking), and developing strong interpersonal relationships and supportive friendships that carry on beyond the scope of the activity itself. This is particularly true for clergy and their spouses, which is useful for those organisations responsible for clergy wellbeing and motivation (such as dioceses and national church organisations).

Ordained bikers identified the group as a safe space for clergy, outside the parish, where "I can be me", whilst spouses of clergy recognized the benefits of spending time with "others like me who understand the pressures of clergy life", beyond the usual (and apparently less effective) clergy groups or clergy spouse support networks. Overall, the motivations for members of the Biker Revs group match those of traditional pilgrims and other biker pilgrims, such as those who participate in the 'Run for the Wall'. Stipendiary clergy, in particular, value the benefits of this close community, where they find the freedom to act as they wish, safe amongst their peers. Finally, the trigger and purpose of the Biker Revs group has achieved its objective–to bring clergy and laity together from across the new diocese, building new relationships and support mechanisms, especially for clergy.

This paper offers substantive and literature-supported analysis of the diversification of motivations, perception of sacred journeys and authentic emotions accompanying pilgrimages; it presents the relationship between pilgrimage, the mission of proclaiming God's word, and a type of religious tourism, in the context of the sense of community as well as time for contact with the sacred; it takes a qualitative ethnographic approach and considers the emerging perspectives of authenticity and credibility; and interprets sacred space through the analysis of its perception and shaping the sense of social 'communion'.

The primary data demonstrates that this topic has future research potential in many areas, including the development of social capital and informal networks that thrive in this context; comparative studies with other biker groups, in the UK and internationally; male spirituality; personality and biking; transformational aspects of pilgrimage; life-changing impacts of biking trips for clergy; the development of clergy support networks and relationships; and the continuing development of the concept of eventization of faith.

Funding: This research was partially funded by retreat grants from The Diocese of Leeds (Anglican).

Acknowledgments: I am grateful to the Biker Revs for their time, to colleagues at the International Conference on Religious Tourism and Pilgrimage in 2017 for their encouragement, and to the three peer reviewers for their helpful feedback.

Conflicts of Interest: The author declares no conflict of interest.

References

Abad-Galzacorta, Marina, and Basagaitz Guereño-Omil. 2016. Pilgrimage as Tourism Experience: The Case of the Ignatian Way. *International Journal of Religious Tourism and Pilgrimage* 4: 48–66.

Alford, Steven, and Suzanne Ferriss. 2006. Classifying Motorcycle Writing: A preliminary consideration. *International Journal of Motorcycle Studies*. Available online: http://ijms.nova.edu/July2006/IJMS_Resources.AlfordFerriss.html (accessed on 9 January 2021).

Béres, Laura. 2012. A Thin Place: Narratives of Space and Place, Celtic Spirituality and Meaning. *Journal of Religion and Spirituality in Social Work: Social Thought* 31: 394–413. [CrossRef]

Bogle, Alison. 2016. Diocese to be known as 'Diocese of Leeds'. Leeds Diocese. April 13. Available online: http://www.westyorkshiredales.anglican.org/content/maps-and-information-about-deaneries-and-parishes (accessed on 9 January 2021).

Cassar, George, and Dane Munro. 2016. Malta: A Differentiated approach to the pilgrim-tourist dichotomy. *International Journal of Religious Tourism and Pilgrimage* 4: 67–78. [CrossRef]

Collins, Michael H. 2000. Christian Forgiveness in Northern Ireland. *Contemporary Review* 276: 235–39.

Corrymeela Community. n.d. Available online: http://www.corrymeela.org/about/our-history (accessed on 9 January 2021).

Diocese of West Yorkshire and the Dales. 2014. Map of the new Diocese of West Yorkshire and the Dales. Available online: http://www.westyorkshiredales.anglican.org/content/maps-and-information-about-deaneries-and-parishes (accessed on 9 January 2021).

Dowson, Ruth, Jabar Yaqub, and Razaq Raj. 2019. *Spiritual and Religious Tourism: Motivations and Management*. Wallingford: CABI.

du Cros, Hilary, and Bob McKercher. 2015. *Cultural Tourism*, 2nd ed. London: Routledge.

Dubisch, Jill. 2004. 'Heartland of America': Memory, Motion and the (Re)construction of History on a Motorcycle Pilgrimage. In *Reframing Pilgrimage: Cultures in Motion*. Edited by Coleman Simon and Eade John. London: Routledge.

Dubisch, Jill. 2008. Sites of Memory, Sites of Sorrow: An American Veterans' Motorcycle Pilgrimage. 2008. In *Shrines and Pilgrimage in the Modern World: New Itineraries into the Sacred*. Edited by Peter J. Margry. Amsterdam: Amsterdam University Press.

Dubisch, Jill, and Michael Winkelman. 2005. *Pilgrimage and Healing*. Tucson: University of Arizona Press.

Esposito, Roberto. 2010. *Communitas: The Origin and Destiny of Community*. Translated by Campbell Timothy. Stanford: Stanford University Press.

Fladmark, Jan M., ed. 1998. *In Search of Heritage: As Pilgrim or Tourist?* Aberdeen: The Robert Gordon University Heritage Library.

Frisk, Henrik, and Stefan Östersjö. 2013. Beyond Validity: Claiming the Legacy of the Artist-Researcher. *Svensk Tidskrift för Musikforskning* 95: 41–63.

Kaell, Hillary. 2016. Notes on Pilgrimage and Pilgrimage Studies. *Practical Matters Journal* 9: 5–14.

Kujawa, Joanna. 2017. Spiritual Tourism as a Quest. *Tourism Management Perspectives* 24: 193–200. [CrossRef]

Leeds Diocese. 2014. Available online: www.leeds.anglican.org (accessed on 9 January 2021).

Leppäkari, Maria, and Kevin Griffin, eds. 2017. *Pilgrimage and Tourism to Holy Cities: Ideological and Management Perspectives*. Wallingford: CABI.

Merkel, Udo, ed. 2015. *Identity Discourses in International Events, Festivals and Spectacles*. Basingstoke: Palgrave Macmillan.

Michalowski, Raymond, and Jill Dubisch. 2001. *Run for the Wall: Remembering Vietnam on a Motorcycle Pilgrimage*. New Brunswick: Rutgers University Press.

Moore, Robert L. 2001. *The Archetype of Initiation: Sacred Space, Ritual Process and Personal Transformation–Lectures and Essays*. Edited by Max J. Havlick Jr. Bloomington: Xlibris Corp.

Norman, Alex. 2011. *Spiritual Tourism: Travel and Religious Practice in Western Society*. London: Bloomsbury Academic.

Pfadenhauer, Michaela. 2010. The Eventization of Faith as a marketing strategy: World Youth Day as an innovative response of the Catholic Church to pluralization. *International Journal of Nonprofit and Voluntary Sector Marketing* 15: 382–94. [CrossRef]

Power, Rosemary. 2006. A Place of Community: 'Celtic Iona and Institutional Religion. *Folklore* 117: 33–53. [CrossRef]

Raj, Razaq, and Kevin A. Griffin, eds. 2015. *Religious Tourism and Pilgrimage Management: An International Perspective*. Wallingford: CABI.

Remsberg, Rich. 2000. *Riders for God: The Story of a Christian Motorcycle Gang*. Urbana: University of Illinois Press.

Scharen, Christian B., ed. 2012. *Explorations in Ecclesiology and Ethnography*. Grand Rapids: William B. Eerdmans Publishing Company.

Scriven, Richard. 2014. Geographies of Pilgrimage: Meaningful Movements and Embodied Mobilities. *Geography Compass* 8: 249–61. [CrossRef]

Shackley, Myra. 2001. *Managing Sacred Sites: Service Provision and Visitor Experience*. London: Continuum.

Silverman, David. 2019. *Interpreting Qualitative Data*, 6th ed. London: Sage Publications.

Stillman, Sean. 2018. *God's Biker: Motorcycles and Misfits*. London: SPCK.

Sturm, Damion. 2015. 'Fluid Spectator-Tourists': Innovative Televisual Technologies, Global Audiences and the 2015 Cricket World Cup. *Communicazioni Sociali* 2015: 230–40.

Symes, Colin. 2009. Early Northumbrian Spirituality: The Fruit of Straitened Circumstances. *Journal of European Baptist Studies* 10: 6–18.

The 59 Club. n.d. Available online: https://www.the59club.co.uk/history-c1d3n (accessed on 9 January 2021).

Turner, Victor, and Edith Turner. 1978. *Image and Pilgrimage in Christian Culture*. New York: Colombia University Press.

Ward, Peter, ed. 2012. *Perspectives in Ecclesiology and Ethnography*. Grand Rapids: William B. Eerdmans Publishing Company.

Wiggen, Todd C. 2019. An Autoethnographic Exploration of Social Identity and Leadership within a Motorcycle Club. *Journal of Motorcycle Studies* 15: 2019.

MDPI
St. Alban-Anlage 66
4052 Basel
Switzerland
Tel. +41 61 683 77 34
Fax +41 61 302 89 18
www.mdpi.com

Religions Editorial Office
E-mail: religions@mdpi.com
www.mdpi.com/journal/religions

www.ingramcontent.com/pod-product-compliance
Lightning Source LLC
LaVergne TN
LVHW070843170726
843515LV00005B/560
* 9 7 8 3 0 3 6 5 1 8 6 7 1 *